Analog Electronics for Measuring Systems

FOCUS SERIES

Series Editor Mireille Mouis

Analog Electronics for Measuring Systems

Davide Bucci

iSTE

WILEY

First published 2017 in Great Britain and the United States by ISTE Ltd and John Wiley & Sons, Inc.

ISTE Ltd
27-37 St George's Road
London SW19 4EU
UK

www.iste.co.uk

John Wiley & Sons, Inc.
111 River Street
Hoboken, NJ 07030
USA

www.wiley.com

Library of Congress Control Number: 2017930069

British Library Cataloguing-in-Publication Data
A CIP record for this book is available from the British Library
ISSN 2051-2481 (Print)
ISSN 2051-249X (Online)
ISBN 978-1-78630-148-2

Contents

Introduction

I.1. Purpose

A measuring system is a coordinated ensemble of different devices allowing a measurement operation through their interaction.

Thanks to its intrinsic flexibility, electronics is a powerful tool available to measurement science. This book is therefore dedicated to the exploration of several recurrent problems in this context, for what concerns the analog part of the measurement chain. We try to follow the usual analog signal path through a general acquisition chain and we describe the elements most frequently found there, with a level of generality sufficient to be useful in different domains (physics, biology…).

Figure I.1 shows the most traditional and general organization of a complete acquisition system from the sensor to the data storage system. Every measurement operation starts with a goal, which is the determination of a quantity (temperature, gas pressure, electric signals from heart beating, etc.). This quantity is called the *measurand*.

The sensor has the role of translating the measurand into an electrical quantity. If needed, the latter is in turn transformed into a voltage by a *signal conditioning system*. Chapter 1 describes the most used classes of sensors along with some classic conditioning strategies.

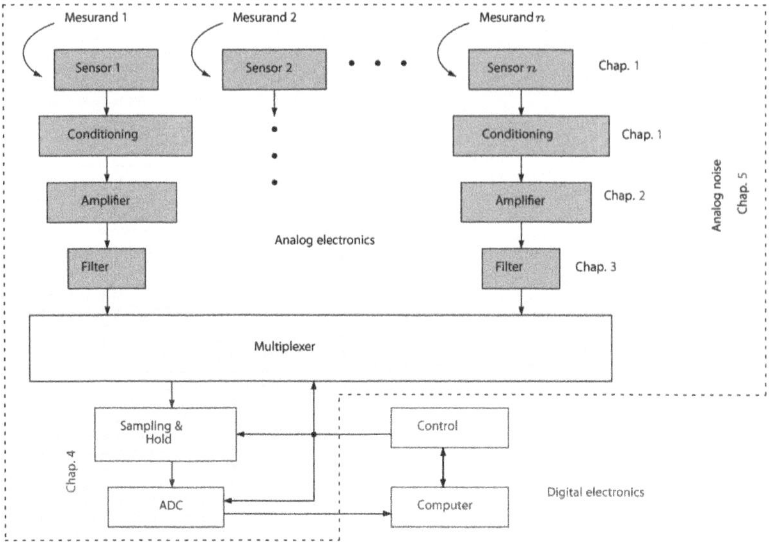

Figure I.1. *Functional organization of a very general acquisition system. Shaded elements are treated in this book, in the chapters indicated*

The output voltage is then *amplified* and *filtered* to obtain amplitudes that are easy to manipulate and reduce as much as possible the noise, which is inevitably present along with the useful signal. Amplifiers (especially differential ones) and filters are, respectively, described in Chapters 2 and 3. In fact, filtering the analog signal has a paramount importance in those situations where a risk of signal aliasing appears. Filters employed in this context usually have a low-pass response and are called *anti-aliasing filters*. The overall quality of a measurement chain depends (even critically in some cases) on the quality of such a filter.

Digital electronics offers a huge range of very advanced signal-processing capabilities. It is very easy, today, to acquire a signal with an analog to digital converter in order to further process it or for storage purposes. The interface between analog and digital worlds is assured by a *sample and hold circuit*, working in tandem with an *analog to digital converter*. Those two devices can be shared among different separate acquisition channels because of a *multiplexer*. This is described in Chapter 4.

Noise is the companion of every analog circuit and the main performance limiting factor. Understanding its origins and behavior is, therefore, a key factor to design high-performance systems. We briefly introduce noise analysis in low-frequency circuits in Chapter 5.

Finally, a *control system* monitors every element of the measurement system, and usually a *computer* manipulates acquired data for storage or visualization. We will not discuss these elements in this book.

In this book, we discuss the analog elements described above to a certain degree of detail: sensors, amplifiers and filters, for low-frequency acquisition systems. We insist that the overall quality of measurements is determined individually by each element through its interaction in the chain. For this reason, when possible, we present some examples, inspired by application notes and literature.

I.2. Prerequisites

This book is addressed to readers with a background in electronic circuits who want to begin to have an idea of the usual problems that arise when designing low-frequency analog circuits that treat the signal coming from a sensor. To limit the overall size of the book, we decided to concentrate on solutions based on discrete devices and integrated circuits (i.e. the specific problems associated with the design of analog integrated circuits will not be addressed). The main prerequisites are:

– AC and DC analysis of circuits, transfer functions and basics of operational amplifiers;

– concepts of power, calculation and interpretation of the root mean square value of a voltage of a current;

– being able to subdivide a complex circuit in more elementary blocks;

– know the most frequently used electronics devices and understand data sheets and technical literature dedicated to real devices;

– basic concepts of signal processing (Fourier transform, sampling Nyquist–Shannon theorem, filtering);

– basic probability and statistical tools (probability density functions, expected values, statistical independence, etc.).

Those prerequisites are addressed in undergraduate electronics courses in most engineering faculties as well as books [MAL 15].

I.3. Scope of the book

When writing a book about engineering, it is somewhat difficult to find the good trade-off between abstraction and practical craftsmanship that together constitute the core of a field such as electronics. We choose to employ maths when necessary (for example while discussing filter synthesis in Chapter 3 or for the noise analysis in Chapter 5), yet we tried to keep the mathematical developments close to the engineering problems and the real-world intuition.

On the other hand, when possible, we present extract from data sheets and technical literature. It should be clear, however, that we do not want to endorse a particular producer or a particular model. We just selected those components that, for a reason or another, may appear to be rather significative of a certain class of devices.

The relation between electric circuits and measurement techniques started very early in the 19th Century and still continues today. This means that:

– an incredibly huge number of solutions are already known for the most disparate measurement situations;

– ready-made low-cost integrated circuits and modules are available, accomplishing wonderfully complex measurement tasks.

Having said that, reading a small book about electronic measurement techniques may seem a futile exercise. Something has to be considered though. First of all, knowing how things work helps when a ready-made solution fails to accomplish its duty. In fact, a culture about analog electronic circuits is useful today more than ever, and culture is no black magic.

Moreover, after all, someone has to do the hard stuff since ready-made solutions do not build themselves alone.

Of course, we live in a society where access to information is widespread and inexpensive. There are of course excellent textbooks in public libraries, but it is also easy and very convenient to browse for technical information on the Internet. However, one must know what to search for and must already have a solid background to critically select the most relevant and meaningful search results from the "noise floor".

Some application notes from the semiconductor industry are wonderfully written and incredibly informative. For example, it is a sheer pleasure to read Jim William's application notes from Linear Technology. They are crystal clear, full of analog wisdom, intellectually honest, informative and fun. In one word, they are terrific. Some of them are explicitly cited among the references. Other resources are simply not worth reading and contain errors or obscure and uninformative descriptions. Particularly dull are those that, instead of producing real original content, just copy/paste information found elsewhere, with minor cosmetic changes.

This book may constitute an useful starting point for deeper investigations.

I.4. Conventions for schematics and voltages

NOTE.– Color versions of the figures in this book (where applicable) have been made available at www.iste.co.uk/bucci/analog.zip.

Figures in this book have been drawn with FidoCadJ, an open source multiplatform program. The symbols employed in this book are the classical symbols for components employed in electronic engineering and should not be ambiguous.

However, a risk of confusion exists for a specific point: we indicate voltages in the figures by means of arrows, whose heads point toward the conventional positive terminal. This is the traditional convention followed, for example, in Italian and French engineering faculties. However, in other places, the opposite convention is followed: be careful if you are not used to this notation.

Finally, when we talk in general of "the voltage of a node of a circuit", the conventional negative term is implicitly supposed to be the reference node. We employ p for the Laplace variable, except when, during filter synthesis, we normalize the frequency. In this case, we indicate it with s.

I.5. Acknowledgments

This book originates from a collection of handouts written for a course in analog electronics and taught to biomedical engineering undergraduate students in GrenobleINP-Phelma. I would have never tried to transform my crude course handouts into this book without the constant encouragement of Dr. Mireille Mouis (IMEP-LAHC), whom I would like to thank very warmly. Those early handouts contained countless issues and errors, which have been pointed out by students, who also had a number of useful suggestions. My colleagues Pr. Laurent Aubard (Grenoble INP-Phelma), Pr. Franco Maddaleno and Dr. Massimo Ortolano (both from Politecnico di Torino) then spotted a lot of errors and suggested highly valuable improvements and corrections on obscure or imprecise points. I am also deeply thankful to Prs. Quentin Rafhay and Irina Ionica (both from GrenobleINP-Phelma/IMEP-LAHC) who carefully read the manuscript and provided a number of precious suggestions. I am also profoundly indebted to Dr. Marc Arques (CEA DTBS Grenoble) for the time he took from his activities to discuss noise and theory of stochastic signals and for providing interesting insights. I would like to thank Sophie Cornu (Grenoble INP-Phelma) who gave a hand, quite literally, for the photographs shown in Figure 1.6.

I tried my best to contact all the copyright owners for the pictures and graphs I reproduced from component datasheets. I would like to thank all those who kindly replied. However, in some cases, my messages were probably never read. If you own the copyright of one of those pictures reproduced here and you are not happy about that, please contact me (or my publisher) and we will collaborate to remove the offending content.

I hope that, thanks to all those who helped me, you now have a document in your hands that has improved from the first versions. Perfection is not of this world, it is quite certain that typos as well more embarrassing errors still linger. The responsibility being solely mine, I apologize in advance. Any feedback and correction will be appreciated.

1

Fundamentals of Sensing and Signal Conditioning

1.1. Introduction

The first element in a classic electronic measuring system is the sensor. Its role is to translate the physical quantity to be measured (called *measurand*) in an electrical quantity of some kind. Clearly, the goal is to obtain knowledge about the physical quantity: this translation should be done in a well-known and reliable way. Sensors are based on a wide range of principles and (our point of view being from the electronics side) we follow the classification proposed in [ASC 03]. In other words, in this chapter we will categorize sensors depending on the electrical quantity at their output: voltage, current, charge, resistance and reactance. This categorization is not the only applicable one, but it allows us to treat signal conditioning at the same time as sensors.

1.2. Voltage generating sensors

1.2.1. *General description*

Several physical phenomena involve the presence of a voltage between two conductors in a specific piece of equipment. That voltage can be related to a particular physical variable. They can thus be exploited to build sensors that can be seen as voltage sources and whose voltage depends on the measurand m. Very often, the electrical representation of a sensor might be a Thévenin-type equivalent circuit including a series impedance, as shown in Figure 1.1.

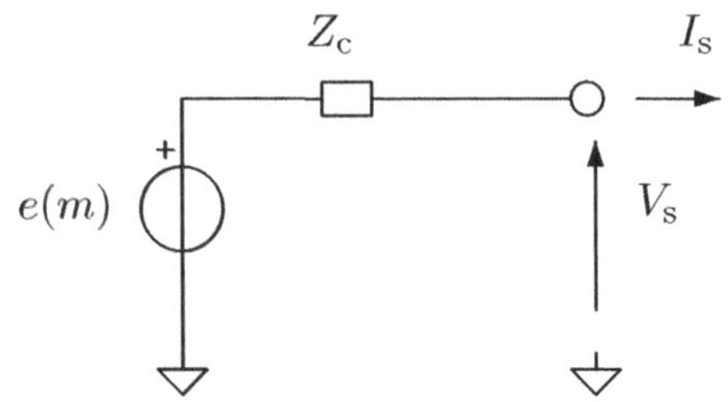

Figure 1.1. *Thévenin representation of a voltage generating sensor*

The open-circuit voltage given by the sensor is $e(m)$ and its relation with m, the measurand, must be known and must not change considerably in time. The internal impedance of the sensor is represented by Z_C and determines the voltage drop between $e(m)$ and V_S when a load is attached, if the current I_S is negligible.

We will proceed by detailing some examples of such sensors, that we judge are the most representative.

1.2.2. *Examples*

1.2.2.1. *Thermocouples*

A thermocouple is a temperature sensor based on the Seebeck effect. This principle, shown in Figure 1.2, is when two junctions between different metallic conductors are kept at different temperatures and a voltage difference can be measured [ASH 76]. This voltage is approximatively proportional to the temperature difference between the two junctions. This effect is a consequence of heat transport in conductors and the Seebeck coefficient is the volume property of each one. The Seebeck voltage, thus, is not generated in the junctions themselves, but on the whole length of the conductor: it is always present, even in a homogeneous circuit, but it only is observable with different conductors spliced together.

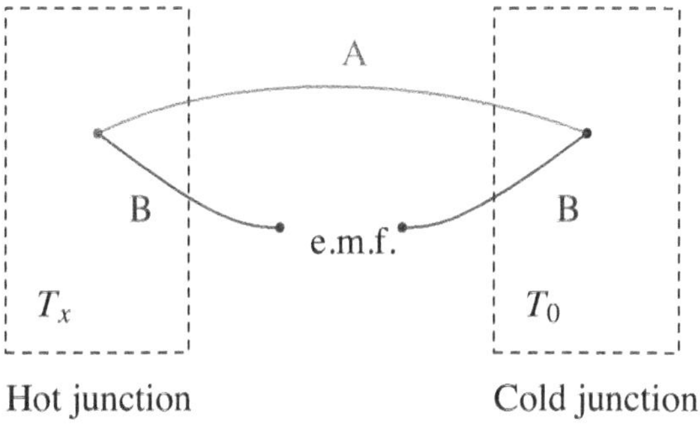

Figure 1.2. *The Seebeck effect in two junctions of different conductors (A and B)*

In the thermocouple's case, we conventionally call "hot junction" and "cold junction" the connections between the two conductors and the temperature measurement is intrinsically differential. If one needs an absolute measurement, the cold junction should be kept at a constant and controlled temperature (for example employing water/ice for 0°C), or a *cold junction compensation circuit* may be used.

From a practical point of view, buckets containing ice and water have long ago been replaced by compensation circuits, which are more compact, much easier to run and less expensive. The idea is to measure the temperature of the cold junction with a physical principle different from the Seebeck effect and subtract its influence from the signal delivered by the thermocouple. The advantage of this method is that the cold junction is much less subjected to extreme temperatures or harsh conditions than the hot junction so the measurement is easy. Among the available strategies, a common solution is to make this subtraction directly to the voltage delivered by the thermocouple. Figure 1.3 shows how it can be done by means of a simple analog circuit.

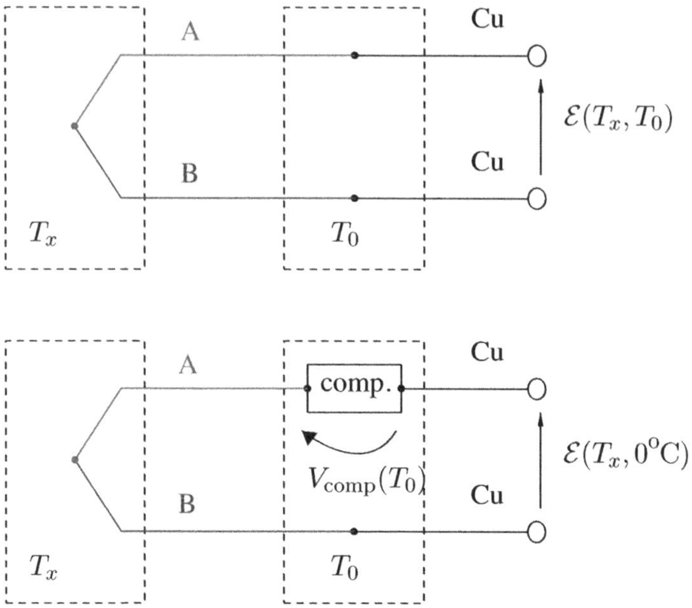

Figure 1.3. *A cold junction compensation of a thermocouple measurement system. Temperature T_0 of the cold joint is measured and translated into a voltage $V_{\text{com}}(T_0)$, substracted from the thermocouple output*

The need of a cold junction compensation circuit entails an increase in complexity of the measurement system. Moreover, the generated voltages are quite small (the sensitivity is around $41\,\mu\text{V}/^{\text{o}}\text{C}$ for a K-type thermocouple at an ambient temperature) and an amplifier is always necessary. However, thermocouples are used in industry very often since they are extremely rugged. They can also work reliably in a wide range of temperatures (from cryogenic temperatures to beyond 1,700°C).

For this reason, compact integrated solutions exist and are sold by microelectronic industries. For example, we cite the AD8494-7 family, an extract of their data sheet is visible in Figure 1.4. Those integrated circuits are able to amplify thermocouple signals while doing an internal compensation of the cold junction. The chip is able to measure its own temperature in order to do the compensation. Of course, this only works if the real cold junction is located very close to the integrated circuit.

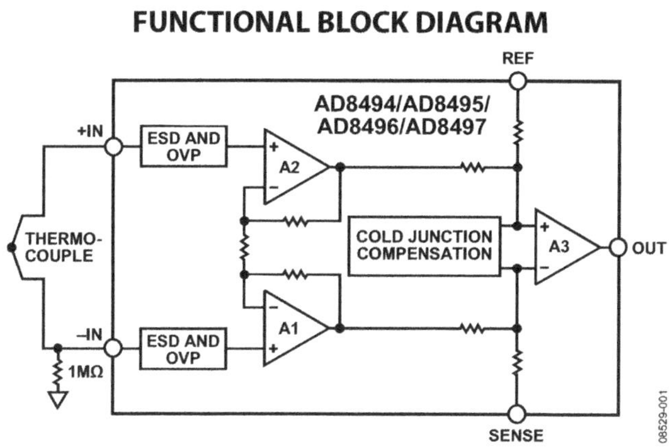

Figure 1.4. *An extract of the data sheet of Analog Devices AD8494-7 family. This device amplifies the thermocouple signal, compensating the cold junction temperature at the same time. ESD and OVP are the electrostatic discharge and over voltage protections for input pins (source: Analog Devices)*

1.2.2.2. *Thermocouple families*

Various kinds of thermocouples are available on the market and are identified by a one letter code. The most frequently used families of thermocouples [ASC 03] are as follows:

– **K** chromel (nickel-chrome)/alumel (aluminum–nickel);

– **E** chromel (nickel-chrome)/constantan (copper–nickel);

– **J** iron/constantan (copper–nickel);

– **N** nicrosil (nickel–chrome–silicon)/nisil (nickel–silicon);

– **T** copper/constantan (copper–nickel);

– **S** platinum–rhodium (10% Rh and 90% Pt, by weight)/platinum;

– **R** platinum–rhodium (13% Rh and 87% Pt, by weight)/platinum;

– **B** platinum–rhodium (30% Rh and 70% Pt, by weight)/platinum–rhodium (6% Rh and 94% Pt, by weight).

In reality, as well as a certain chemical composition, this letter indicates a well-specified relation between temperature and voltage. The relation should conform to the ideal one within a known tolerance. The definition of the models and the tolerances is usually given by institutes of standards.

Typical thermocouple behaviors are shown in Figure 1.5. Some thermocouples families (such as E or J) deliver a higher output voltage but are affected by a stronger nonlinearity or they can only be used in a smaller range of temperatures. Other thermocouples (such as S, R and B) deliver a smaller signal but are renowned for their long-term stability and capability of measuring very high temperatures.

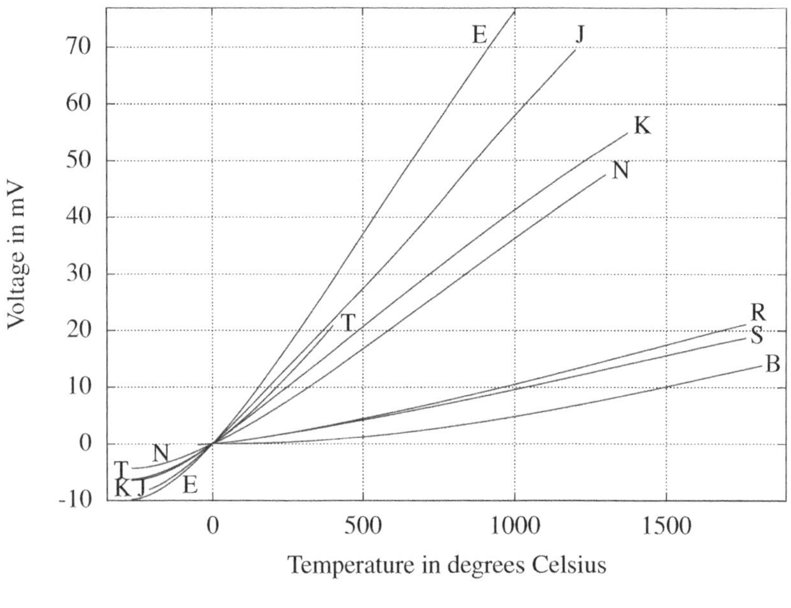

Figure 1.5. *Relation between temperature and output voltage for some common thermocouple types. The cold junction is kept at* $0\,^{o}\mathrm{C}$*. Data from [NIS 90]*

In order to take into account the nonlinearities in a wide temperature range, the law $\mathcal{E}(T_x, 0^{o}\mathrm{C})$ is very often written as a polynomial equation to which some exponential terms are added for certain thermocouple families:

$$\mathcal{E}(T_x, 0^{\text{o}}\text{C}) = \sum_{i=0}^{N} c_i T_x^i + a_0 \text{e}^{a_1 (T_x - a_2)^2} \qquad [1.1]$$

The National Institute of Standards and Technology in the United States publishes a catalog of tables of coefficients to be adopted for the thermocouples mentioned above [NIS 90]. For instance, Table 1.1 shows the coefficients useful for the K-type thermocouple in several temperature ranges.

	range: -270 to $0\,^{\text{o}}$C	range: 0 to 1,372 ^{o}C
c_0	$0.\ 000000000000 \times 10^{+0}$	$-0.\ 176004136860 \times 10^{-1}$
c_1	$0.\ 394501280250 \times 10^{-1}$	$0.\ 389212049750 \times 10^{-1}$
c_2	$0.\ 236223735980 \times 10^{-4}$	$0.\ 185587700320 \times 10^{-4}$
c_3	$-0.\ 328589067840 \times 10^{-6}$	$-0.\ 994575928740 \times 10^{-7}$
c_4	$-0.\ 499048287770 \times 10^{-8}$	$0.\ 318409457190 \times 10^{-9}$
c_5	$-0.\ 675090591730 \times 10^{-10}$	$-0.\ 560728448890 \times 10^{-12}$
c_6	$-0.\ 574103274280 \times 10^{-12}$	$0.\ 560750590590 \times 10^{-15}$
c_7	$-0.\ 310888728940 \times 10^{-14}$	$-0.\ 320207200030 \times 10^{-18}$
c_8	$-0.\ 104516093650 \times 10^{-16}$	$0.\ 971511471520 \times 10^{-22}$
c_9	$-0.\ 198892668780 \times 10^{-19}$	$-0.\ 121047212750 \times 10^{-25}$
c_{10}	$-0.\ 163226974860 \times 10^{-22}$	
	Exponential	
a_0	$0.\ 118597600000 \times 10^{+0}$	
a_1	$-0.\ 118343200000 \times 10^{-3}$	
a_2	$0.\ 126968600000 \times 10^{+3}$	

Table 1.1. *Table of coefficients to be used in equation* [1.1] *to calculate the output voltage of a K-type thermocouple. Measurement units of the coefficients are such that the output voltage is in millivolt and the temperature is in degrees Celsius. Data published by [NIS 90]*

Very often, for small temperature ranges (or when high accuracy is not sought), only the first linear term is taken into account in the calculations. This term is called sensitivity for short. For a K-type thermocouple, in the temperature range caught between 0 and 100 ^{o}C, an approximation yielding an accuracy of a few degrees Celsius employs only a linear term $S \approx 41\ \mu\mathrm{V}/^{\mathrm{o}}\mathrm{C}$.

1.2.2.3. *pH measurement*

Determining the pH of a solution is one of the most frequent characterizations useful with chemicals. It consists of measuring the acidity or basicity and it can be done using litmus paper, which dipped in a solution changes its color depending on the pH.

Another technique is based on the use of glass electrodes specifically built for this function. They contain buffer solutions that exchange H^+ ions with the solution being tested via a semipermeable glass membrane. Figure 1.6 shows photographs of a typical pH electrode of this kind (Sentek P14/S7). Due to the electrochemical reactions, the sensor behaves like a battery and a DC voltage is obtained, proportional to the pH to be measured. The proportionality constant depends on the exact configuration of the electrode, but a typical sensitivity is around $-60\,\mathrm{mV}$ for pH unit.

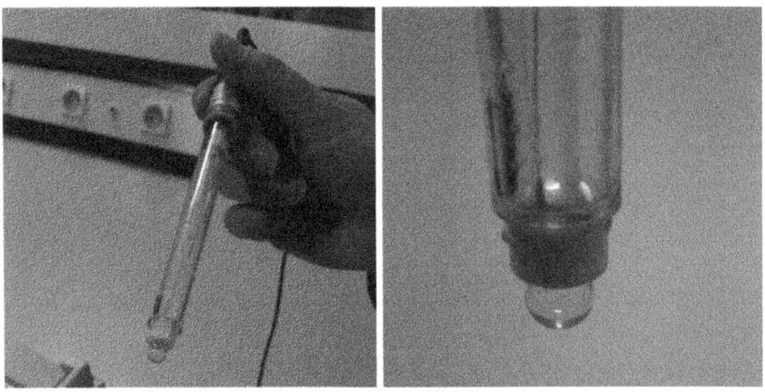

Figure 1.6. *Photographs of the Sentek P14/S7 electrode for pH measurements*

One of the difficulties of measuring pH is that the series impedance of the glass electrode (called Z_c in the equivalent circuit in Figure 1.1) is often very high, of the order of a few megaohms. For this reason, such probes may be equipped with an onboard amplifier very close to the measurement electrodes.

A second difficulty is the strong temperature effect on the proportionality between the pH and the output voltage. Often, an automatic compensation must be done by measuring the temperature of the solution at the same time as the pH measurement. For example, in the Campbell Scientific CSIM-11 probe, the sensitivity changes by $-0.2\,\mathrm{mV/pH/^oC}$. This means that at $20\,^o\mathrm{C}$ the slope is around $-58\,\mathrm{mV/pH}$, while it is equal to $-59\,\mathrm{mV/pH}$ at $25\,^o\mathrm{C}$, then reaching $-60\,\mathrm{mV/pH}$ at $30\,^o\mathrm{C}$. For precise measurements, a calibration is required at regular intervals using well-known buffered solutions.

1.3. Current generating sensors

1.3.1. *General description*

When a physical action induces the generation of charge carriers in a material, this phenomenon may result in a variation of the current flowing in the device. The electrical output of a sensor exploiting this principle is thus a current. Some examples are as follows:

– radiation-induced ionization effects;

– carriers generation by photoelectrical effect.

As the information on the measurand is carried by a current, it is natural to adopt an equivalent circuit representation that is a Norton equivalent, as represented in Figure 1.7. The measurand m is translated to a certain current $i(m)$, which flows through the terminals of the generator. The internal admittance of the sensor is given by Y_C. Of course, the model shown in Figure 1.7 is greatly simplified and might not be able to represent the real behavior of the sensor when V_S exceeds particular limits: nonlinearity often lurks around the corner. In the following section, we will discuss some examples of sensors of this kind.

1.3.2. *Examples*

1.3.2.1. *Photomultipliers*

When a photon impinges on a conductor and if the energy carried is high enough, it can extract an electron that becomes a free carrier. This effect is called external photoelectrical effect and it has been discovered in the 19th Century and studied by A. Einstein in one of his seminal papers [EIN 05] published in his *Annus Mirabilis*. By exploiting this principle, a light sensor can be built by applying an electrical field to move the carriers, while monitoring the current circulating in the system. The current will be in fact proportional to the flux of photons in the unit of time related to the light intensity. An important condition is, however, that the energy of the photons is high enough to enable the photoelectrical effect.

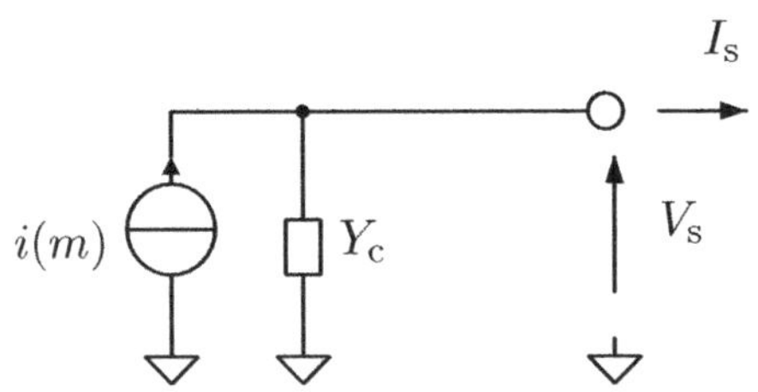

Figure 1.7. *Norton equivalent circuit of a current generating sensor*

Figure 1.8 shows a schematic view of a device that can be used to measure light via the photoelectric effect. An important value quantifying the overall quality of the device is called the *quantum efficiency* η, i.e. the ratio between the number of generated electrons and the impinging photons. A second parameter (related to Y) is the *sensitivity* R, the ratio between the generated current I_{ph} and the impinging light power P. The measurement unit of this parameter is thus A/W.

In practice, generated currents are often quite small and handling them might become tricky. The presence of a current amplification internal to the sensor itself might simplify the task of detecting very low optical intensities. This is done in photomultipliers by adding a number of intermediate electrodes (dynodes) to multiply the number of electrons by exploiting the secondary emission of electrons. A single photon thus results in a significant number of electrons because of the amplification process. This principle is represented in Figure 1.9. Dynodes are often biased by a resistive network from the photomultiplier power supply rails.

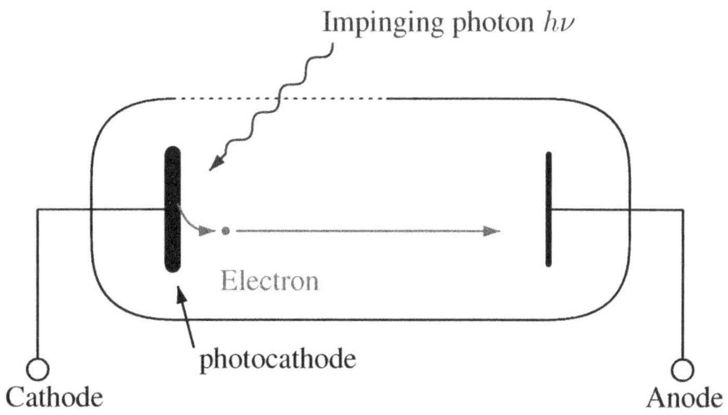

Figure 1.8. *Working principle of a sensor based on the photoelectrical effect*

An extract of the data sheet of a multiplier tube can be found in Figure 1.10. It detects light in the visible range of wavelengths and it is thus sensitive between 300 and 850 nm. The photocathode is followed by 10 gain stages, and the electron gain depends exponentially on the bias voltage, adjusting the sensitivity of the sensor.

Photomultipliers are able to generate an event for just one photon impinging, but not all photons will trigger it. They are relatively bulky, fragile and they need high voltages, yet photomultipliers are still in use today especially when a very high sensitivity and low noise are not an option.

1.3.2.2. *Photodiodes*

A semiconductor is characterized by the presence of energy bands where carriers move reacting to an electric field. Electrons in the valence band might be brought in the conduction band following the absorption of a photon having enough energy $E = h\nu$. This only happens if $E > E_g$, where E_g is the energy gap of the semiconductor. This is called *internal photoelectric effect* in a similar way to what is described in section 1.3.2.1. An absorbed photon involves the generation of a pair of carriers, an electron in the conduction band and a hole in the valence band.

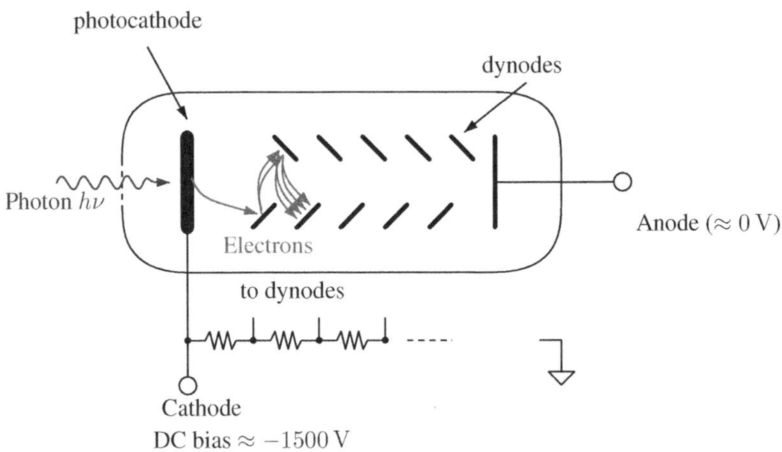

Figure 1.9. *Working principle of a photomultiplier*

In a PN junction, the internal field allows the separation of the electron/hole pair which results in a certain current. This current is proportional to the number of generated carriers per unit of time. That is in turn proportional to the number of absorbed photon flux in the junction, hence the absorbed optical power. The proportionality constant R is called responsivity or sensitivity and it is defined exactly as in photomultipliers:

$$R = \frac{I_{ph}}{P}, \quad [1.2]$$

where I_{ph} is the photocurrent and P the optical power impinging on the device. The electrical symbol of a photodiode is shown in Figure 1.11. The figure also shows how the current/voltage characteristics of the diode is vertically translated when the junction receives a certain flux of photons and a photocurrent appears.

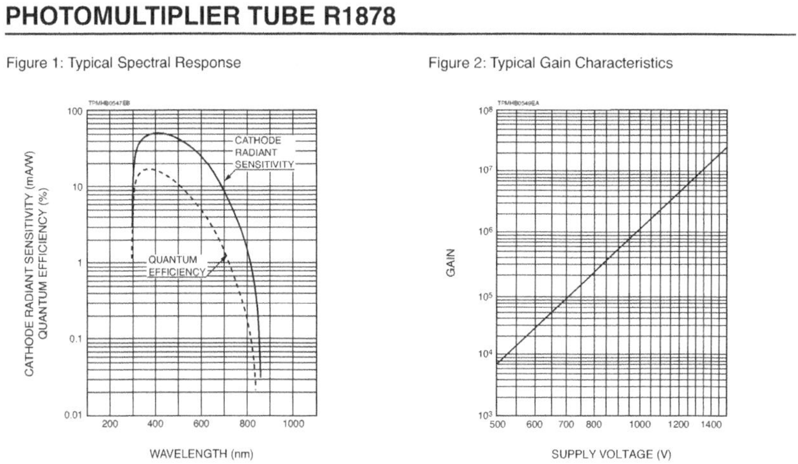

Figure 1.10. *Extract of the data sheet of an R1878 photomultiplier tube. On the left, the quantum efficiency as well as the responsivity of the photocathode is represented versus the wavelength. On the right, the gain versus the bias voltage is represented. Courtesy of Hamamatsu Photonics K.K*

Hugely different applications exist, ranging from light power measurements to colorimetry, detection of telecom signals in optical fibers and so on.

Figure 1.12 shows an extract of the data sheet of a silicon photodiode. The silicon band gap is $E_g = 1.12\,\text{eV}$ and this means that photons having a wavelength longer than $1,100\,\text{nm}$ are practically not absorbed since they do not have enough energy to generate an electron/hole pair. The sensitivity is thus dramatically reduced when the impinging light has a longer wavelength. Nonlinear effects still exist above that limit, involving the simultaneous absorption of two or three photons but only play a role for a very intense light.

In the equivalent circuit shown in Figure 1.11 there are two current generators. The first one I_{ph} is associated with the photocurrent obtained by the internal photo-electric effect. The second generator I_s represents the dark current due to the minority carriers drift across the junction when it is reversely biased. We then have the internal resistance (ranging from several megaohms to a few gigaohms) as well as the junction capacitance. In some applications, the capacitance C_d might become a limiting factor for the detection speed or the stability of the conditioning circuit.

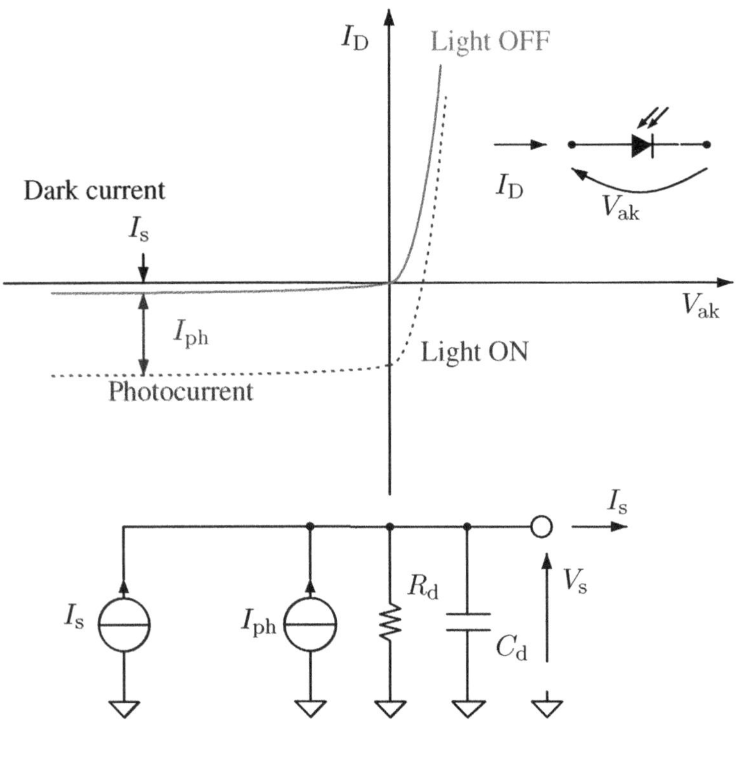

Figure 1.11. *Electrical symbol of a photodiode, its typical I/V characteristics in obscurity and with light impinging as well as its equivalent circuit*

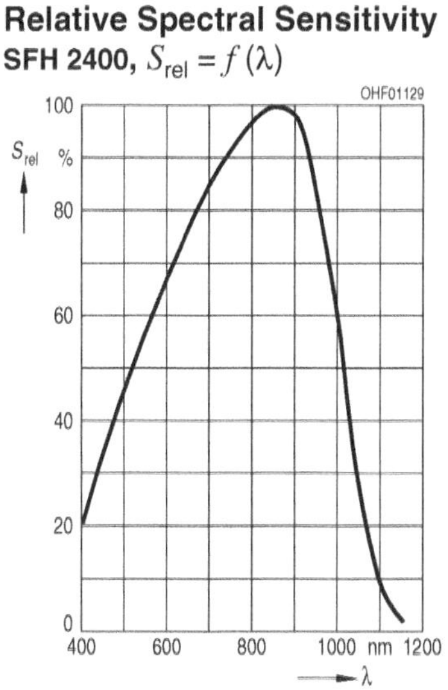

Figure 1.12. *Sensitivity versus the wavelength of the SFH2400 silicon photodiode. Courtesy of Osram-OS*

A particular class of devices are the avalanche photodiodes. In this case, the carriers generated in the junction by a photon absorption are multiplied because of an avalanche effect in a strongly reversely biased device. Like photomultipliers, there is thus an internal gain of the photodiode, which is useful for the detection of very low light intensities.

1.3.3. *Conditioning circuits*

A conditioning circuit converts the electrical signal at the output of the sensor into a voltage. Therefore, in this section we deal with current to voltage converters. Several solutions exist, here we discuss the simplest and most current ones.

1.3.3.1. *One resistance*

Sometimes, just one resistance in a strategic place of the circuit can do the job. This solution is often used with photomultipliers and an example can be seen in Figure 1.13, where the voltage signal $e(t)$ is obtained via the resistance R.

A problem that might appear is that the voltage at the output of the circuit affects the biasing of the sensor, thus producing some nonlinear effects if it is large enough. In fact, this solution is also sometimes used with photodiodes, as shown in Figure 1.14; however, a nonlinearity appears when the voltage $e(t)$ at the circuit's output is close to the threshold voltage of the diode. In this condition, the photodiode becomes directly biased; the total current in the device is therefore the algebraic sum of the photocurrent and the forward current. Therefore, the voltage at the terminals of the resistance is no longer in a linear relation with the power of light impinging on the device. The method is, however, quite simple and employed in radio frequency applications, where $50\,\Omega$ is used, eventually by superposing a reverse bias in order to reduce nonlinearity with strong light intensities.

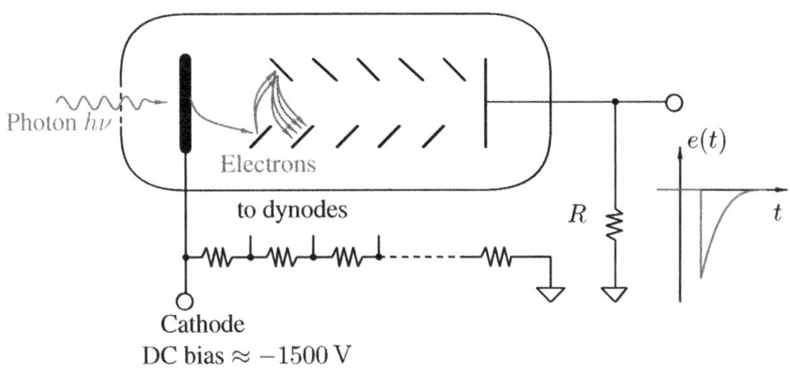

Figure 1.13. *Conditioning of the current signal coming from a photomultiplier via the resistance R*

1.3.3.2. *Transresistance amplifier*

In section 1.3.3.1, we saw that one of the problems associated with signal conditioning using a simple resistance is that the sensor itself is subjected to the same voltage obtained at the output of the conditioning circuit. In some cases, the biasing of the sensor might be distributed, giving rise to nonlinear effects. It would be better to make sure that V_s

is always close to zero, and the current I_s gives rise to a proportional voltage *elsewhere* in the circuit, on nodes different from the terminals of the sensor.

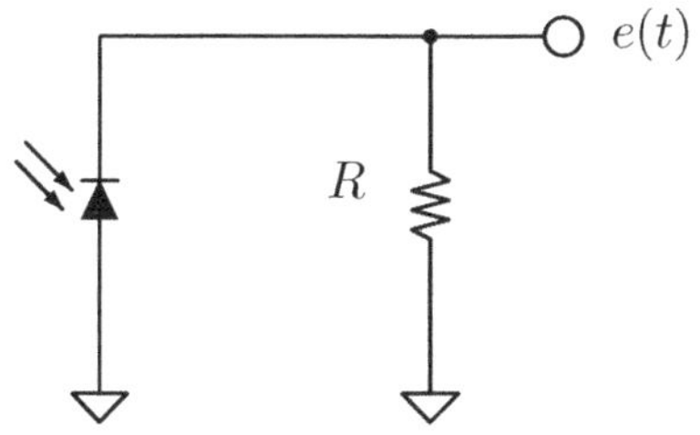

Figure 1.14. *Conditioning of the current signal coming from a photodiode via the resistance R*

A classical solution is the transresistance circuit shown in Figure 1.15, often called "current to voltage converter". An advantage of the circuit is that, because of the properties of the operational amplifier, the voltage across the sensor is very close to zero, thus reducing the influence of the Y_c admittance in the equivalent circuit shown in Figure 1.7. In fact, in the circuit the feedback is provided by the resistance R and, by supposing that the operational amplifier is ideal, it can be shown that the output voltage of the circuit is:

$$V_e = -RI_e \qquad [1.3]$$

If the goal is to obtain a very high transresistance, for example to detect very small currents, R should be very high. Values of resistance up to several gigaohms can be found in catalog, but they tend to be very expensive and have to be handled with care. The circuit shown in Figure 1.16 represents a classical workaround, avoiding extreme values of a single resistance. Circuit analysis leads to:

$$V_e = -\frac{R_1R_2 + R_2R_3 + R_1R_3}{R_3} I_e \qquad [1.4]$$

The proportionality term might thus be seen as equivalent to a very high value resistance. There are, however, some problems associated with this circuit, namely the greatly increased noise if compared to the single resistance, as well as an increase in the offset (which can be interpreted as a sort of "noise in DC").

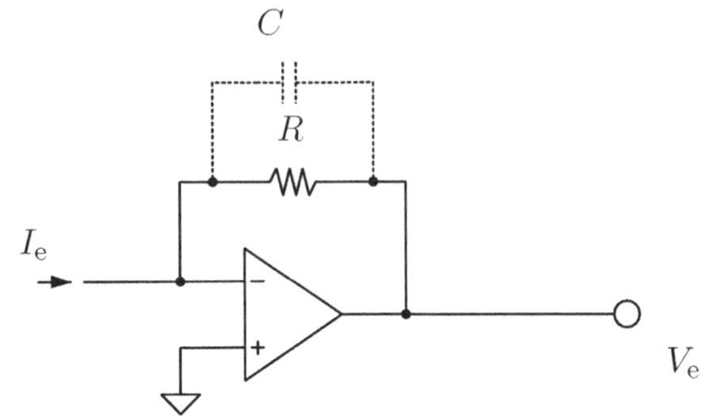

Figure 1.15. *Classical circuit of a transresistance amplifier built around an operational amplifier*

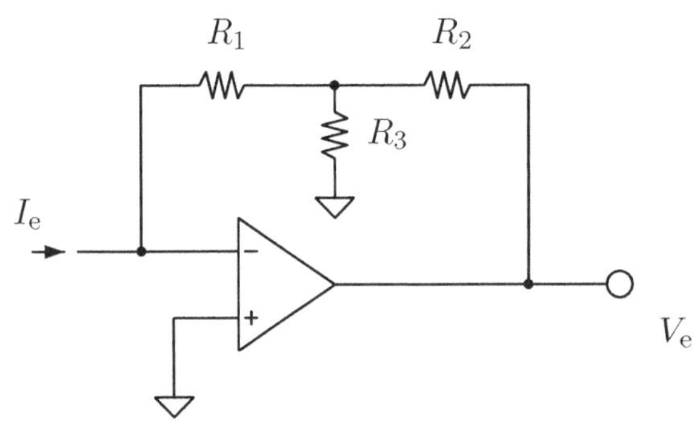

Figure 1.16. *Use of a T-bridge feedback circuit in the current to voltage converter*

In the circuits shown in Figures 1.15 and 1.16, some key points should be taken into account:

– Stability: sensors may possess a relatively large output capacitance, as we have seen for photodiodes (the C_d capacitor in the equivalent circuit). In this case, real-world operational amplifiers might experience instability in the circuits proposed above. This is very well known and it is discussed in application notes [WAN 05] or books [FRA 15] on the subject. A common solution is to place an appropriate capacitor C, in parallel with the resistance R shown in Figure 1.15.

– If small currents should be detected, the bias currents of operational amplifiers have to be very small too. MOS-based devices are available on the market with astonishingly low bias currents. For example, the LMC6001 declares in the data sheets a bias current of 25 fA at 25 °C. As a second example, we cite the LMP7721, with the guaranteed maximum of 20 fA at 25 °C, with a typical value of 3 fA. This is rather breathtaking if we observe that 1 fA represents something like 6000 electrons per second. Application note from Burr-Brown/Texas Instruments [BUR 94], discusses some specific caveats of low current measurements with their OPA128.

– In a circuit, when currents of less of a nanoampere are to be treated, moist and dirt on the printed circuit board (PCB) play a role, which is no longer negligible. Specific techniques must be adopted in the most delicate parts of the circuits (guard rings, shielding, teflon sockets or insulators, etc.). Prepare yourself to wear gloves to handle the most delicate components and work as cleanly as possible.

1.4. Charge generating sensors

1.4.1. *General description*

Some sensors are based on the change in dielectric polarization on a dielectric in response to external stimuli. For example:

– In piezoelectric materials such as quartz crystals or specific ceramics and polymers, a mechanical deformation due to an applied force leads to the appearance of an electric field. Therefore, if the sensor is inserted in a circuit, charges move to counterbalance the field.

– Small-scale rearrangements of the structure of some dielectrics occur when the temperature changes, giving rise to an electric field. This effect is called pyroelectricity. Examples include triglycine sulfate crystals and tourmaline.

In those situations, the measurand is translated into a certain charge unbalance from the equilibrium condition (so the charge is not *actually* generated inside the sensor). Figure 1.17 represents two different ways of modeling this family of sensors. The first circuit on the left is a Thévenin equivalent, which is completed sometimes by a resistance in parallel with the C_0 capacitor, to take into account internal sensor losses. The second equivalent circuit on the right is a Norton equivalence. The choice of the former or the latter equivalent circuit is often a matter of convenience, but both reflect the impossibility of performing DC measurements.

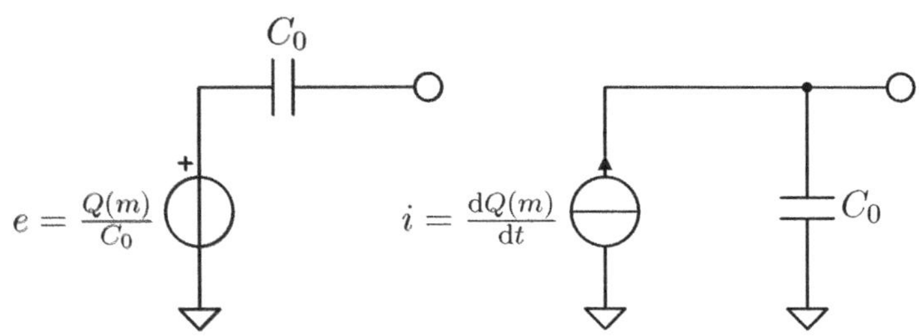

Figure 1.17. *Equivalent circuits for modeling charge generating sensors*

1.4.2. *Examples*

1.4.2.1. *Pyroelectric sensors for infrared sensing*

Probably the most widespread use of pyroelectric sensors is for thermal infrared detection. Motion detectors inside buildings are often based on the detection of sources of heat by a pyroelectric sensor. They are often used to switch on the light when a person enters a certain area or to signal an intrusion. Figure 1.18 shows a typical sensor able to detect infrared radiation whose wavelengths are caught between 5 and 12 μm (in the thermal range).

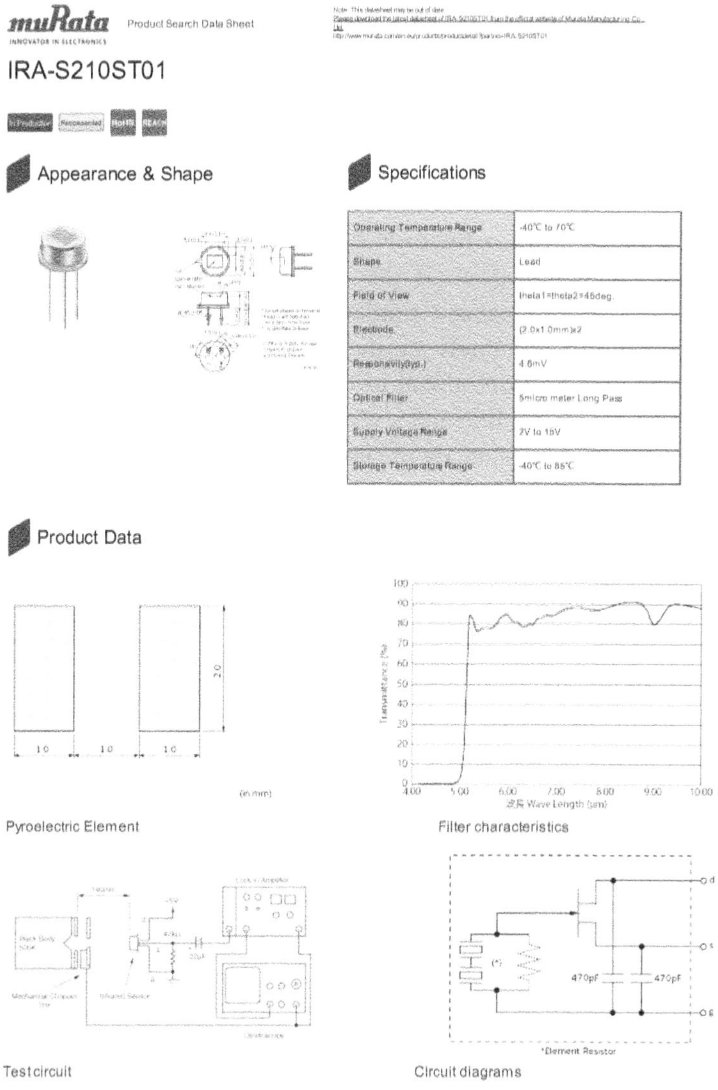

muRata INNOVATOR IN ELECTRONICS Product Search Data Sheet

IRA-S210ST01

Appearance & Shape

Specifications

Operating Temperature Range	-40°C to 70°C
Shape	Lead
Field of View	theta1=theta2=45deg.
Electrode	(2.0x1.0mm)x2
Responsivity(typ.)	4.6mV
Optical Filter	5micro meter Long Pass
Supply Voltage Range	2V to 15V
Storage Temperature Range	-40°C to 85°C

Product Data

Pyroelectric Element

Filter characteristics

Test circuit

Circuit diagrams

Figure 1.18. *Extracts from the data sheet of a IRA-S210ST01 pyroelectric sensor. Courtesy of Murata*

It is interesting to note the presence of a field effect transistor inside the sensor's package to amplify signals coming from the sensor. A complete motion detector system usually couples the pyroelectric sensor with an amplifier/threshold circuit, a plastic Fresnel lens and often a timer.

1.4.2.2. *Piezoelectric sensors*

The application of a force on a piezoelectric material leads to a change in its polarization. If the material is applied between two electrodes in an arrangement similar to a capacitor, a voltage appears between the two terminals of the sensor. This is due to a non-equilibrium charge proportional to the applied mechanical constraint. This principle is reversible and a wide variety of sensors and actuators based on this phenomenon are of everyday use.

Here are some examples: microphones, accelerometers, displacement sensors, etc. We can also cite an interesting musical application: the percussion sensor in the pads of electronic drums is often piezoelectric.

1.4.3. *Conditioning*

A variety of techniques exists for the conditioning of electrical signals coming from charge-based sensors (see, for example, [KAR 00] for a discussion specific to piezoelectric sensors).

A useful idea is that if a capacitor is storing a certain charge, there is a resulting voltage on its terminals and vice versa. This can be exploited at the sensor itself via a very high impedance amplifier. By using the Thévenin equivalent circuit shown in Figure 1.17, it can be seen that the internal capacitor C_0 is charged by a certain voltage, which can be measured if some care is taken not to discharge the capacitor too much during measurements.

The second possibility is to use the circuit shown in Figure 1.19, where an operational amplifier is used as an integrator for the current provided by the sensor, yielding a voltage proportional to the charge.

Some circuit analysis done by assuming an ideal operational amplifier leads to the observation that no current is circulating in capacitors C_0 (sensor capacitance) and C_c (cable/connection capacitance), since in this case, the sensor is connected to a virtual ground. In the same way, the resistance R_c representing the losses plays no role. All the provided charge is stored in the C capacitor, in the feedback loop around the operational amplifier. If the capacitor is fully discharged at the beginning of the measurement, the voltage at the output of the circuit will thus be:

$$V_o = -\frac{Q(m)}{C} \qquad [1.5]$$

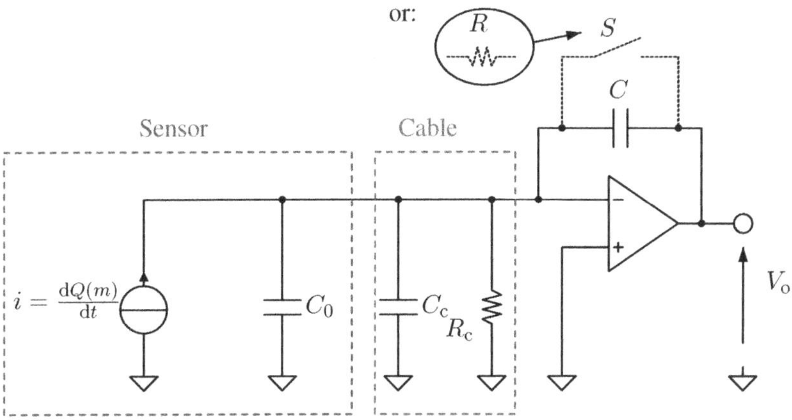

Figure 1.19. *Conditioning circuit useful for charge-based sensors. C_c and R_c are the total capacitance and parasitic shunt resistance of the connection cables*

Something that must be taken into account in this circuit is that there is no DC feedback path (i.e. in the absence of S and R in Figure 1.19). Once the circuit is switched on, the C capacitor will slowly charge itself with the bias currents of the operational amplifier. This means that an error source is present and, after a while, V_o becomes close to the power supply rails of the amplifier, which then saturates.

A first strategy to avoid this phenomenon is to place a switch (more realistically, a field effect transistor operating as a switch in response to a control signal) in parallel with the C capacitor. The purpose of the switch is to discharge it at the beginning of each measurement.

A second solution is to put a resistance R in parallel with C. A Laplace domain analysis yields:

$$V_{\mathrm{o}}(p) = -\frac{Q(p)}{C}\frac{pRC}{1+pRC}, \quad [1.6]$$

which is after all quite similar to equation [1.5], where we have a term that has a high-pass behavior, with a $-3\,\mathrm{dB}$ cutoff frequency:

$$f_{\mathrm{c}} = \frac{1}{2\pi RC}. \quad [1.7]$$

As long as R and C are chosen in such a way that the unattenuated band is compatible with the useful signal, this circuit will be reliable.

1.5. Resistive sensors

A lot of physical phenomena are closely related to the electrical resistance of a conductor or a semiconductor. In fact, every times something modifies the following characteristics, there is a change in the measured resistance:

– carrier mobility (temperature, constraint, magnetic field);

– carrier density (temperature, light absorption);

– geometrical dimensions (constraint, cursor displacement).

A wide class of sensors exploits those effects. Their equivalent circuit is shown in Figure 1.20. We thus present some examples in the following sections.

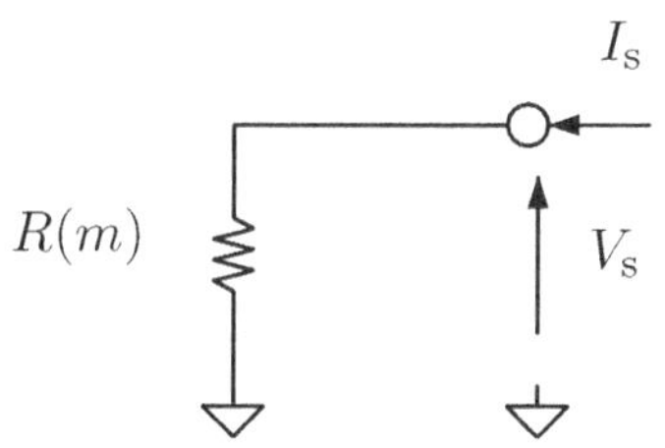

Figure 1.20. *Equivalent circuit of a resistive sensor: a resistance whose value depends on measurand m*

1.5.1. *Examples*

1.5.1.1. *Light-dependent resistors*

If we consider a generic semiconductor, resistivity strongly depends on the carrier concentration (electrons and holes) participating in the conduction. In an intrinsic semiconductor, carrier density is quite low, and the resistivity is comparatively high. If the semiconductor is subject to a photon flux, if their energy is sufficient to generate electron/hole pairs, a significant reduction of the resistance can be observed.

It is thus possible to obtain a device whose resistance strongly depends on the light intensity to which it is subjected, also called a photo-resistance or light-dependent resistors (LDR). Devices based on cadmium sulfide (CdS cells) are widely used for a lot of applications involving visible light since they are simple to make, low in cost and rugged. Figure 1.21 shows the resistance versus light intensity plot of a Luna Optoelectronics NORPS-12 LDR. It can be noted that the device might be used on quite a large dynamic range, such as 4 decades, its resistance in complete darkness being of a few megaohms[1]

1 The light intensity is measured in old foot-candles units; the correct unit in the SI standards is the lux. The conversion factor is 1 foot-candle = 10.764 lux. The multipliers are also non-standard.

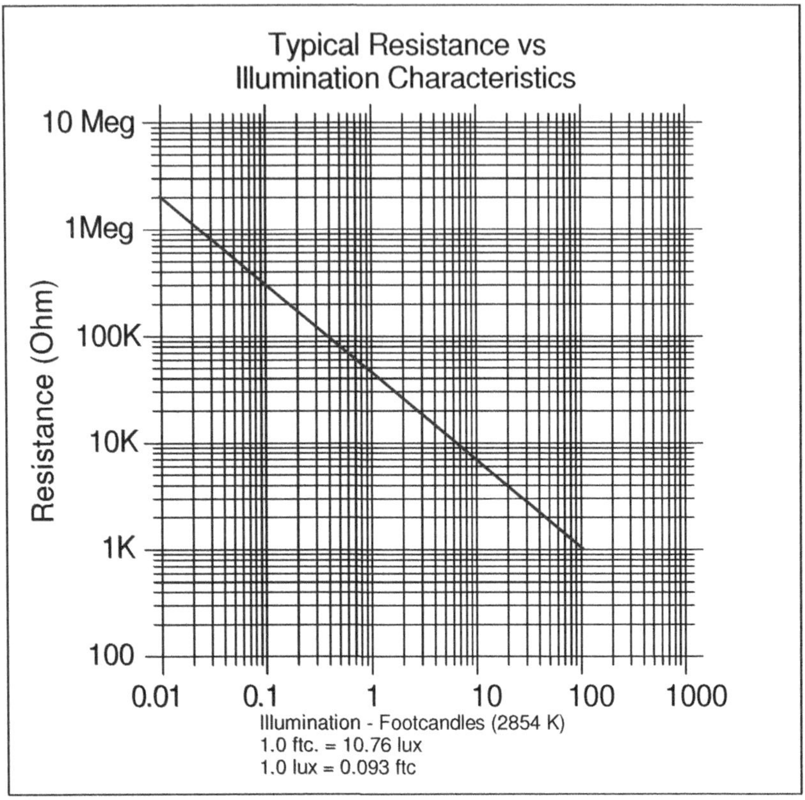

Figure 1.21. *An extract of the data sheet of the NORPS-12 LDR built by Luna Optoelectronics (source: Luna Oploelectronics)*

1.5.1.2. *Platinum temperature probe "Pt100"*

A "Pt100" probe is formed by a given length of platinum wire on a glass or ceramic insulating support. Its resistance R is $R_0 = 100\,\Omega$ at a temperature of $t_x = 0^{\text{o}}\text{C}$ and the temperature dependence is well represented by the following equation:

$$R(t_x) = R_0(1 + \alpha t_x) \qquad [1.8]$$

where $\alpha = 3.85 \times 10^{-3}/^{\text{o}}\text{C}^{-1}$, the *average* nominal temperature coefficient in the range between 0 to $100\,^{\text{o}}\text{C}$. The linear model of equation [1.8] allows us to obtain an error of less than $0.5\,^{\text{o}}\text{C}$ in that interval of temperatures. As we saw about thermocouples in section 1.2.2.2, more complex models can be employed. The IEC 60751 norm (from 2008, previously called IEC 751) specifies the relation between temperature and electrical resistance and defines two tolerance classes (A and B) depending on the desired precision.

A similar principle is used on variants such as the Pt500 or Pt1000, the only notable difference being that in equation [1.8] R_0 is, respectively, 500 and 1,000 Ω.

Figure 1.22 shows an extract of a data sheet common to several temperature probes from Omega engineering, including Pt100 probes as well as thermocouples.

1.5.1.3. *Strain gages*

Strain gages are based on the observation that a conductor subjected to a mechanical constraint varies its electrical resistance. For example, we observe that a homogeneous conductor wire pulled with a certain force F will increase its length l while reducing slightly its section S. The strain is defined as the relative change in length of the conductor due to the applied force: $\frac{\Delta l}{l}$. This phenomenon is reversible, at least if the force F does not exceed the limits for the elastic deformation of the wire. The electrical resistance of the wire can be calculated via the well-known equation:

$$R = \rho\frac{l}{S}, \quad [1.9]$$

where ρ is the resistivity of the material employed, which also depends on the strain applied (piezoresistive effect). In any case, for small variations, the change in R is usually represented as follows[2]:

$$\frac{\Delta R}{R} = K\frac{\Delta l}{l} \quad [1.10]$$

2 The length l is not the only parameter that plays a role in equation [1.9]. Therefore, K includes all the effects related to the change in resistivity and the section of the conductor.

where the proportionality coefficient K is called gauge factor and might vary between 2 and 5 for most metals and 50 and 200 (in module) for semiconductors. In most cases, it is the variation of the resistivity ρ that yields the most important contribution to the resistance change in equation [1.9], and not the change in the geometrical dimensions.

Pt100 RTD Probes & Thermocouples

for Industrial Applications

- Pt100 in class A, 1/3 DIN and 1/10 DIN available
- Thermocouples J, K, T, E & L with accuracy to IEC class 1
- Temperature range -100°C to +400°C
- Standard closed end sensors or airstream models available
- Standard lengths up to 200mm, optional lengths Of 201-500mm
- Diameter 2mm, 3mm or 6mm
- A range of mounting threads
- All Pt100 sensors have a 4-wire connection
- A broad range of lead wire materials available

Omega temperature probes for industrial applications are produced with a wide variety of mechanical attachment systems suitable for a range of process conditions.

They may be specified to your precise requirements by the sensor type in either a closed-end or airstream version, accuracy, probe length, different threads, wire material in shielded and unshielded versions.

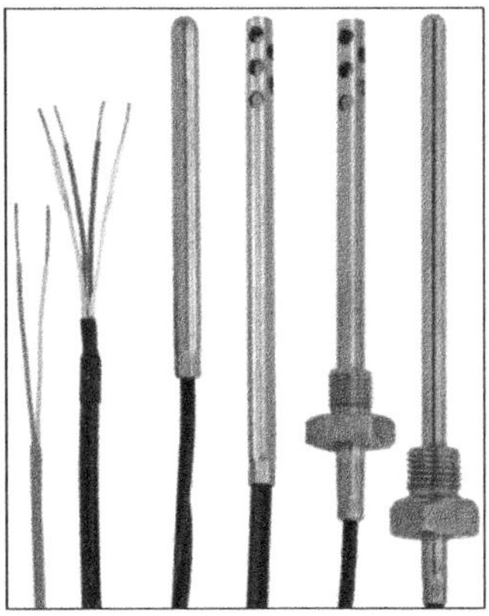

AVAILABLE VERSIONS:

Probe Type	Style	Accuracy	Diameter	Probe Length	Mounting Bush	Lead Wire Insulation	Lead Wire Length
P-Pt100 J-Thermocouple J K-Thermocouple K T-Thermocouple T L-Thermocouple L E-Thermocouple E N-Thermocouple N	M-Closed end L-Airstream Probe H-Hand Held Probe EH-Hand-held probe with penetration tip	P-Pt100(4-wire): A-Class A 1/3-1/3 DIN B 1/10-1/10 DIN B Thermocouples: 1-Class 1 (standard)	6-6mm 3-3mm 2-2mm	5-200mm 200-500mm	O-Plain stem M6-M6x1 pitch M8-M8x1 pitch M10-M10x1 pitch G1/8-1/8" BSP Pl. G1/4-1/4" BSP Pl. G1/2-1/2" BSP Pl.	P-PVC S-Silicone Rubber T-Teflon (FEP) G-Glass Fibre PS-PVC Shielded TS-Teflon Shielded GS-Glass Shielded	XM-Cable length in metres

www.omega.co.uk +44 (0)161 777 6611

Figure 1.22. *A family of platinum temperature sensors and thermocouples © Copyright Omega Engineering, Inc. All rights reserved. Reproduced with the permission of Omega Engineering, Inc. Norwalk, CT 06854 www.omega.com*

1.5.2. *Caveats*

When one needs to measure a resistance, several factors should be carefully analyzed:

– resistance is always associated with heat generated by the Joule effect. The electrical power employed for the measurement must therefore be controlled to avoid self-heating which might induce an error on the measurements;

– the lead wires do not always have a negligible resistance;

– sensors in an industrial environment might operate close to electrically noisy machines, giving rise to electromagnetic compatibility issues;

– electrical insulation might be a delicate issue when dealing with harsh environments (high temperatures, aggressive chemicals, high voltages, radioactivity, etc.)

1.5.3. *Signal conditioning: measuring the total resistance*

Two different situations must be considered:

– information about the measurement is carried by the total resistance $R(m)$;

– information is carried by a variation of the resistance around a certain mean value R_0, so that: $R = R_0 + \Delta R(m)$.

While they are not mutually exclusive, the two cases address different problems and therefore we treat them separately. In this section, we discuss the first situation, where the information about the measurand m is translated by the sensor in a resistance value $R(m)$ to be determined. For example, it is the case for the platinum sensor described in section 1.5.1.2. A simple solution might be to excite the sensor with a known current I_{a}: the voltage V_{m} obtained at the output is hence related to the excitation current by Ohm's law:

$$V_{\mathrm{a}} = R(m) I_{\mathrm{m}} \qquad [1.11]$$

In reality, the sensor might be placed quite far from the conditioning circuit and the lead wire resistances R_f might play a non-negligible role in the measurement, as shown in Figure 1.23 on the left. What can be accessed in this configuration is just the resistance $R(m) + 2R_\mathrm{f}$. In fact, R_f might be difficult to keep stable and depends on a lot of factors (ageing, temperature, nature of the conductors, splices, connectors, etc.). This kind of measurement is called "two wire measurement" to be distinguished from the more complex "four-wire measurement", which we are about to describe and that is depicted in Figure 1.23 on the right. Most of the cases, the voltmeter resistance R_i might be sufficiently high to be considered infinite. With electronic voltmeters, it is not uncommon for R_i to range well above $10\,\mathrm{M}\Omega$.

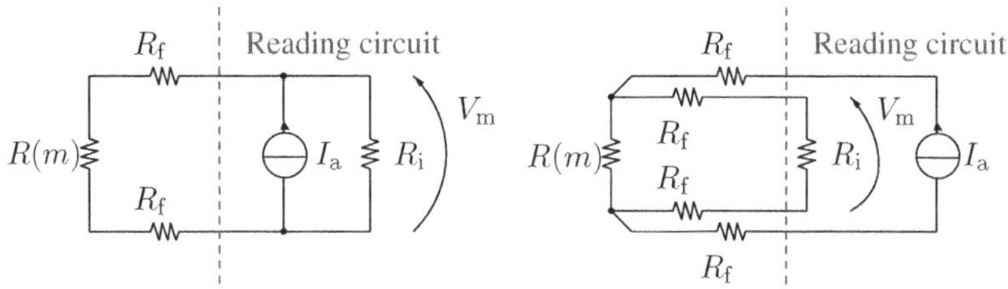

Figure 1.23. *Conditioning strategies for a resistive sensor, measure of the total resistance $R(m)$. On the left: 2-wire measurement. On the right: 4-wire (Kelvin) measurement. R_f is due to cables and connections, R_i is the internal resistance of the voltmeter. For a color version of this figure, see ww.iste.co.uk/bucci/analog.zip*

The idea at the background of the 4-wire technique (also called "Kelvin contact") is that the wires used to measure voltage carry almost no current, whereas the wires carrying the excitation current I_m are not involved in the voltage measurement. The measured voltage V_m thus becomes virtually independent from R_f, allowing to determine $R(m)$ very precisely.

A less expensive variant of this technique involving only three wires, as shown in Figure 1.24, is widely employed in industrial applications. In fact, as long as the lead wire resistance remains approximatively the same for each of the three connections (which can be done if the three wires are kept together in the same cable), it is possible to estimate it at each measurement and compensate it fairly accurately. The resistance between terminals 1 and 3 is:

$$R_{1-3} = 2R_f + R(m), \qquad [1.12]$$

whereas resistance R_{1-2}, measured between terminals 1 and 2, is simply twice R_f. Therefore:

$$R_{1-3} - R_{1-2} = 2R_f + R(m) - 2R_f = R(m) \qquad [1.13]$$

which is the resistance we want to measure.

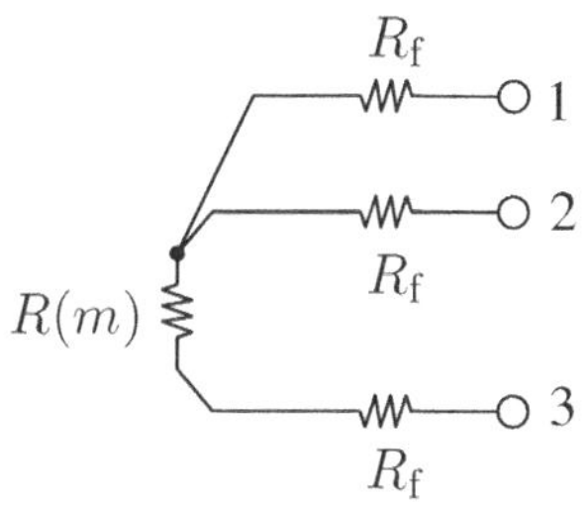

Figure 1.24. *Three-wire conditioning technique for compensating lead wire resistance*

1.5.4. *Measuring a resistance variation: the Wheatstone bridge*

When the goal is to measure a resistance variation around an equilibrium point (this is often the case when dealing with strain gages discussed in section 1.5.1.3), a classic circuit is the Wheatstone bridge, as shown in Figure 1.25. It traces its origins in the first half of the 19th Century and was originally employed to measure an unknown resistance by carefully adjusting other (known) resistances. The principle is to measure the voltage difference between nodes A and B in the bridge circuit. The voltage V_{AB} is equal to zero only for a perfectly balanced bridge, condition that corresponds to the following equation:

$$R_3R_2 = R_1R_4. \qquad [1.14]$$

In order to check whether or not this condition is satisfied, we must measure V_{AB} with precision, as long as we are able to detect a zero. That was technically feasible even at the time when the bridge was invented: precisely measuring a voltage was a real challenge. Nowadays, however, we have wonderfully linear instruments and it is usually far more convenient to measure V_{AB} and to explain its relation with the values of the resistances. This is the path we will follow in the rest of this section.

Two kinds of excitation are usually adopted:

– constant voltage excitation, where the bridge is connected to a voltage source E. The current flowing in the bridge varies during the measure because the total bridge resistance changes. The output voltage V_{AB} is:

$$V_{AB} = \frac{R_2R_3 - R_1R_4}{(R_1 + R_2)(R_3 + R_4)}E \qquad [1.15]$$

– constant current excitation, where the current flowing in the bridge is kept stable to a certain value I_a. The output voltage is as follows:

$$V_{AB} = \frac{R_2R_3 - R_1R_4}{R_1 + R_2 + R_3 + R_4}I_a. \qquad [1.16]$$

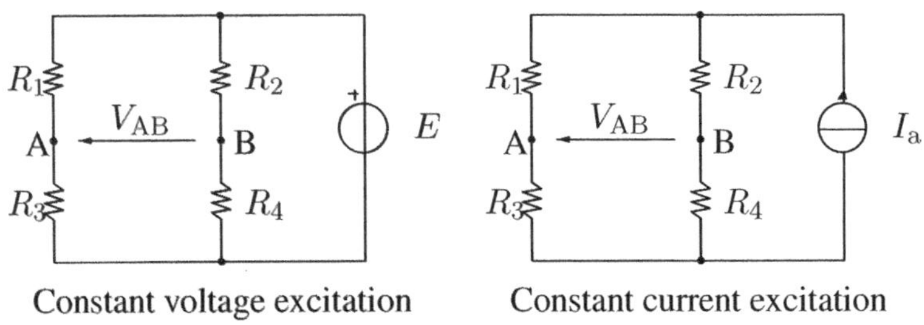

Figure 1.25. *Wheatstone bridges, constant voltage and current excitations*

Despite (or thanks to) its simplicity, the Wheatstone bridge, and its close relatives, is one of the most elegant solutions to a surprisingly wide range of problems [WIL 90]. In practical measurement situations, different variants are exploited, depending on the number of sensing elements in the bridge. Bridge variants exist for 3 or 4 wire connections of the sensing resistance (Kelvin double bridge, also known as Thomson bridge). In the following sections, however, we will only discuss the simplest configurations.

1.5.4.1. *A single variable element*

In the bridge shown in Figure 1.25, just one of the four resistances is a variable, for example R_2. We may write:

$$\begin{cases} R_2 = R_0 + \Delta R_2(m) \\ R_1 = R_3 = R_4 = R_0 \end{cases} . \qquad [1.17]$$

With a constant voltage excitation, by exploiting equation [1.15], we find:

$$V_{\mathrm{AB}} = \frac{\Delta R_2(m)}{R_0} \times \frac{1}{1 + \frac{\Delta R_2(m)}{2R_0}} \times \frac{E}{4}. \qquad [1.18]$$

It can be noted that V_{AB} voltage is nonlinearly dependent from $\Delta R_2(m)$ and this might be a disadvantage in some situations. However, if $\Delta R_2(m)$ is small enough, the following simplification can be carried out:

$$V_{\mathrm{AB}} \approx \frac{\Delta R_2(m)}{R_0} \times \frac{E}{4}. \qquad [1.19]$$

For a constant current excitation of the bridge, we obtain:

$$V_{\mathrm{AB}} = \Delta R_2(m) \times \frac{1}{1 + \frac{\Delta R_2(m)}{4R_0}} \times \frac{I_{\mathrm{a}}}{4} \approx \Delta R_2 \frac{I_{\mathrm{a}}}{4}, \qquad [1.20]$$

which once again reflects a nonlinear relation between V_{AB} and ΔR_2 and the linearization is allowable when the bridge unbalance is small. For the constant current excitation, however, the nonlinear term is twice as small than for a constant voltage excitation.

1.5.4.2. *Two variable elements*

We insert two variable elements in the bridge as follows:

$$\begin{cases} R_1 = R_0 + \Delta R_1(m) \\ R_2 = R_0 + \Delta R_2(m) \\ R_3 = R_4 = R_0 \end{cases} \qquad [1.21]$$

The non-null voltage V_{AB} can thus be calculated:

– constant voltage excitation:

$$V_{\mathrm{AB}} = \frac{\Delta R_2(m) - \Delta R_1(m)}{R_0} \times \frac{1}{1 + \frac{\Delta R_2(m)+\Delta R_1(m)}{2R_0}} \times \frac{E}{4} \qquad [1.22]$$

– constant current excitation:

$$V_{\mathrm{AB}} = [\Delta R_2(m) - \Delta R_1(m)] \times \frac{1}{1 + \frac{\Delta R_2(m)+\Delta R_1(m)}{4R_0}} \times \frac{I_{\mathrm{a}}}{4} \qquad [1.23]$$

We can see that both expressions obtained above might be simplified if a certain symmetry is respected. In fact, it $\Delta R_2 = -\Delta R_1 = \Delta R(m)$ (the so-called "push-pull" configuration), we obtain:

– constant voltage:

$$V_{\mathrm{AB}} = \frac{\Delta R(m)}{R_0} \times \frac{E}{2} \qquad [1.24]$$

– constant current:

$$V_{\mathrm{AB}} = \Delta R(m)\frac{I_{\mathrm{a}}}{2}. \qquad [1.25]$$

An interesting point is that now the latter expressions are perfectly linear and the sensitivity is twice the case where one sensing element is present in the bridge. Very often, this configuration is adopted in cases where symmetrical resistance variations are due to a carefully tailored mechanical symmetrical arrangement.

1.5.4.3. *Four variable elements*

We consider a symmetric situation similar to the "push pull" strategy seen in the previous section, but extending the same concept to four variable resistances:

$$\begin{cases} R_1 = R_0 - \Delta R(m) \\ R_2 = R_0 + \Delta R(m) \\ R_3 = R_0 + \Delta R(m) \\ R_4 = R_0 - \Delta R(m) \end{cases} . \quad [1.26]$$

The unbalance voltage V_{AB} is as follows:

– constant voltage excitation:

$$V_{\mathrm{AB}} = \frac{\Delta R}{R_0} E, \quad [1.27]$$

– constant current excitation:

$$V_{\mathrm{AB}} = \Delta R I_{\mathrm{a}}. \quad [1.28]$$

Because of the symmetry, the expressions are linear.

1.5.4.4. *Example: strain gages and Wheatstone bridge*

The Wheatstone bridge configuration with four variable resistances is a widespread solution with strain gages. In Figure 1.26, we show a photograph of a sensor built with a strain gage glued to a spring (a "load cell"). Similar devices often constitute the heart of small weighing scales. It is very convenient to employ four gauges exploiting the compressive and tensile strains on the spring so that they work in a push–pull configuration as in equation [1.26].

The "sensitivity" parameter is usually specified in data sheets and indicates the ratio between the differential output voltage of the Wheatstone bridge contained in the device and the constant voltage excitation, when the full-scale weight (maximum capacity) is measured. For example, with a sensitivity of $1.8\,\mathrm{mV/V}$, if one excites the sensor with $3.3\,\mathrm{V}$, the full-scale differential output V_{AB} is $5.94\,\mathrm{mV}$ for a model with a maximum capacity of $1\,\mathrm{kg}$ such as the one shown in Figure 1.26.

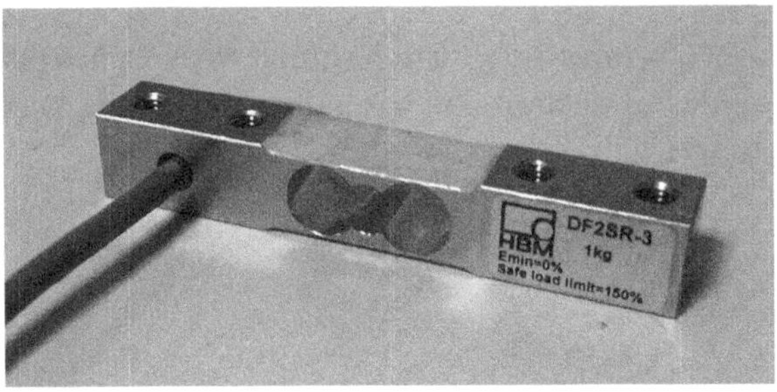

Figure 1.26. *Photograph of the load cell DF2SR-3 from HBM. The strain gages are glued to a spring, whose deformation is translated into a differential voltage*

1.6. Reactive sensors

A vast family of reactive sensors exist. Those devices translate the measurand into a variation of capacity or inductance. For example, a coil embedded in the floor may detect a car at the entrance of a car park Another important application of an inductive linear displacement sensor is the linear variable differential transformer. It is a transformer, where the core can be moved to change the coupling between windings. The input coil is fed with a constant-amplitude AC voltage. Two output coils (connected with a 180° phase) deliver an output AC voltage whose amplitude depends on the position of the core with respect to the coils. Conversely, there is a vast range of proximity or position sensors working on a capacitive principle.

We will not describe this class of sensors in much detail. We will just present some points that may be considered:

– measurement of a reactance involves employing AC signals, not always sinusoidal, but in any case always varying in time;

– if the reactance is put in an oscillator, its changes may be detected through the output frequency variation, easily measurable with digital counters;

– many variants of the Wheatstone bridge exist, adapted to reactive elements and working in AC (Maxwell bridge, De Sauty bridge, etc...).

1.7. Conclusion

In this chapter, several families of different electrical sensors have been described. Our descriptions were oriented toward their representation in an electronics circuit. We have, therefore, focused rather on the electrical characteristics than on working principles. Thus, we have treated at the same time the conditioning circuits adopted for each kind of sensor. We tried to follow a practical approach and several examples from data sheets have been discussed.

2

Amplification and Amplifiers

2.1. Introduction

In Chapter 1, we described a certain number of sensors, as well as the conditioning circuit used to obtain a voltage as an electrical representation of the measurand. Now, this voltage should be somehow treated: in most cases, it should be amplified, and often filtered. This chapter is devoted to the amplification of low-frequency signals and it is particularly focused on circuits based on operational amplifiers. We will therefore begin by briefly describing the working principles of operational amplifiers: in particular, we will focus on some parameters specified in the data sheets quantifying their limits and non-idealities. We will describe here the so-called voltage-feedback operational amplifier (often just called operational amplifier), frequently adopted in low-frequency circuits. A different element, the current-feedback operational amplifier, bears some resemblance to it, but as its use is more specific, it will not be described here. We will then give an overview of differential amplifiers, in particular the instrumentation amplifiers. The name of those circuits reflects their widespread use in instrumentation... The end of the chapter will be devoted to insulation amplifiers, which are very useful when security or electromagnetic compatibility issues are of primary importance.

There are a lot of very good textbooks that develop the matter presented here in detail. One of them is, of course [ASC 03], also cited in the previous chapters. We also recommend [FRA 15], which is very comprehensive and presents some advanced matters.

2.2. Introduction to operational amplifiers

2.2.1. *The operational amplifier as a differential amplifier*

First of all, an operational amplifier is an electronic circuit with two inputs and one output. It aims to be as close as possible to a *differential* amplifier with a very high gain. Figure 2.1 shows an idealized circuit representation of what we expect from an operational amplifier: it takes the voltage difference ϵ measured between the non-inverting "+" and the inverting "-" inputs, it amplifies it and the amplified output voltage V_o is now referred to the reference node:

$$V_o = A_d \epsilon \qquad [2.1]$$

The input is differential and the output is single ended. In practice, the differential gain A_d is very high, but in real circuits its exact value strongly depends on the frequency and a variety of other factors.

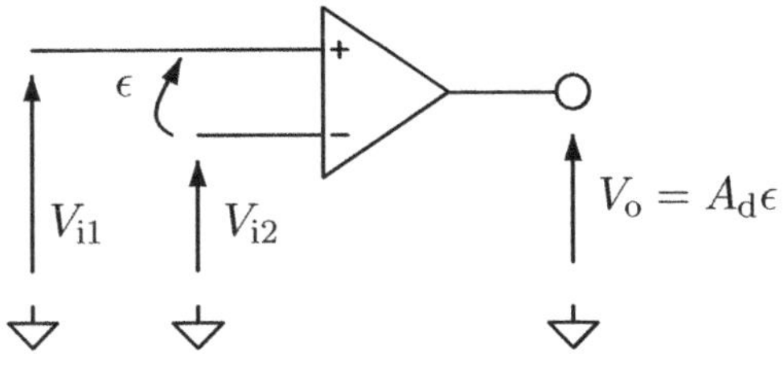

Figure 2.1. *An operational amplifier as a differential amplifier*

Usually, an operational amplifier is drawn as shown in Figure 2.1, where the two inputs and the output are represented[1].

1 While the situation shown in Figure 2.1 is quite common in textbooks, it is evidently a strongly idealized one, or in any case it cannot be complete. In fact, *how can the operational refer its output to the reference node if it does not have any other connection*

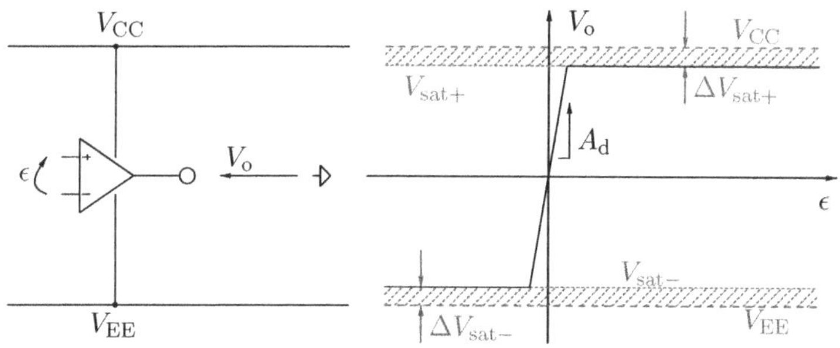

Figure 2.2. *On the left: an operational amplifier with explicit representation of the power supply rails. On the right, a graphical representation of the (idealized) output characteristics*

There is, of course, an important point that has been left out: the power supply rails V_{CC} and V_{EE}, as shown in Figure 2.2 on the left. The same picture depicts a more realistic model of the operational amplifier, by taking into account the saturation: no signal can (at least in ordinary cases) exceed the power supply rails in an operational amplifier circuit. Note that in the picture we represented the power supply rails are symmetrical with respect to the reference potential. With A_d usually ranging between 10^5 and 10^6 (100–120 dB) at DC, if one wants to exploit the linear part of the characteristics, the only way is to use negative feedback. In other words, one must provide a link between the output and the input so that the output can be adjusted by the amplifier to ensure that the differential voltage ϵ at the inputs is always in the linear range of the amplifier.

2.2.2. *Modeling ideal operational amplifiers*

Since the differential gain A_d is so high and moreover related to a lot of factors, we can say that *an operational amplifier is a circuit (often*

---

to it? This means that something must have been left out in the drawing and, for example, a lot of SPICE models for commercial operational amplifiers contain an artificial internal connection to the reference node. Be careful with SPICE simulations: believing that a result is accurate only because it comes from a computer is usually a good way to seek for catastrophes.

quite complex) optimized to be used as a linear block, by means of a feedback network. This feedback network is employed to obtain from the circuit performances that do not depend too much on the gain Ad, as long as it is large enough. A very useful model allows us to simplify the calculations on the circuits by following two rules:

– **R1:** The amplifier is able to measure a differential voltage without perturbing the circuit: no current flows in the inputs.

– **R2:** The differential gain A_d is considered infinite. Therefore, the only admissible situation where V_o is limited is $\epsilon = 0$. We may formulate that by saying that the amplifier does whatever it can with its output, so that its inputs remain at the same voltage.

We note that this way of seeing things requires the presence of a negative feedback: otherwise, the output cannot affect the inputs. We also remark that this simple model does not distinguish between the inverting and non-inverting inputs, while we know that they cannot be exchanged in a real circuit: saturation inevitably occurs if they are confused.

The rules associated with ideal operational amplifiers allow us to understand the basic behaviors of a circuit. However, many key characteristics may only be deduced by taking into account more realistic models. In the following sections, we will therefore briefly describe some of the most relevant limitations of real operational amplifiers. Knowledge of their influence in a circuit allows us to understand the data sheets and select the best product in the huge catalogs proposed by the semiconductor companies.

2.3. Limitations of real operational amplifiers

2.3.1. *Saturation and rail-to-rail operational amplifiers*

In Figure 2.2, we note that the output voltage of the amplifier is bounded inside a range defined by $V_{\mathrm{sat}+}$ and $V_{\mathrm{sat}-}$. One significant figure of merit of the amplifier is the difference between the saturation voltages and the power supply rails ($\Delta V_{\mathrm{sat}+}$ and $\Delta V_{\mathrm{sat}-}$), reflecting the capability of the circuit to work close to V_{CC} and V_{EE}. In classic integrated operational amplifiers, ΔV_{sat+} and ΔV_{sat-} were between 1 and 2 V. Today, we are faced with the widespread use of wireless devices containing batteries, as well as the general trend in digital circuits of reducing power supply voltage. Such a margin would represent a huge reduction of the available dynamic range. In fact, it is not uncommon to seek high-performance analog circuits with a single $V_{CC} - V_{EE} = 3.3\,\text{V}$ supply or even less. For this reason, a class of operational amplifiers (called rail to rail) has been optimized to make sure that $\Delta V_{sat\pm}$ do not exceed 100 mV in the operating conditions.

What we described above relates particularly to the output section of the operational amplifier, but something similar also happens at the inputs: most amplifiers do not work well if the voltages at their inputs are too close to V_{CC} and V_{EE}. Some of the modern rail-to-rail amplifiers, however, tolerate both inputs being slightly above V_{CC} or below V_{EE}, adding flexibility with low voltage supplies.

2.3.2. *Input offset*

For several reasons (mainly some small asymmetries in the fabrication process), when the voltage applied to the two inputs of a real operational amplifier is equal, the output voltage is not zero as predicted by equation [2.1]. In fact, the very high value of the differential DC gain A_d will probably make sure that the asymmetry is exaggerated such that output is saturated, either at V_{sat+} or V_{sat-}. A small DC voltage, called the *offset voltage*, should therefore be applied between the inputs in order that the output is no longer in this condition. The offset voltage can range from a few microvolts in precision operational amplifiers to several millivolts. This effect being static, it is only relevant to those circuits whose bandwidth includes DC. For ultralow offsets (below $50\,\mu\text{V}$), a square wave modulation/synchronous demodulation technique is employed in the so-called auto-zero (or chopper stabilized) amplifiers.

When feedback is present, the presence of offset changes Rule 2 in such a way that the difference between input voltages is no longer null, but equal to the offset voltage. External nulling can often be performed via an external adjustable resistive network.

2.3.3. *Common mode rejection ratio*

In an ideal differential amplifier, the output voltage depends only on the voltage difference between the two inputs, that we called ϵ in Figure 2.1. In a real device, this is not completely true and the average of the two inputs voltages (V_{i1} and V_{i2}) plays a small role. In other words, by supposing that only this error term is present, equation [2.1] should be corrected as follows:

$$V_{\mathrm{o}} = A_{\mathrm{d}}\epsilon + A_{\mathrm{cm}}V_{\mathrm{cm}} \qquad [2.2]$$

where V_{cm} is the so-called common mode, i.e. the arithmetic average of the voltages at the two inputs of the amplifier, each one referred to the reference node. To summarize:

$$\begin{cases} \epsilon = V_{\mathrm{i1}} - V_{\mathrm{i2}} \\ V_{\mathrm{cm}} = \frac{V_{\mathrm{i1}}+V_{\mathrm{i2}}}{2} \end{cases} \qquad [2.3]$$

A new term of gain, namely A_{cm}, the common mode gain, appears. A good differential amplifier (and thus a good operational amplifier) should make sure that A_{d} is much greater than A_{cm}. To quantify this, the data sheets report the *common mode rejection ratio* (CMRR) in decibels, defined as follows:

$$\mathrm{CMRR} = 20\log\frac{A_{\mathrm{d}}}{A_{\mathrm{cm}}} \qquad [2.4]$$

where the logarithm is base 10. Typical figures range between 80 and 120 dB.

2.3.4. *Bias currents*

In section 2.2.2, the first rule states that no current flows in the inputs of an ideal operational amplifier. In real circuits, things are different: *some* current must flow to in the inputs, as the input transistors (usually a differential pair) have to be biased correctly. It is desirable to keep it as small as possible and operational amplifiers have been vastly optimized in this regard: currents in the picoamper range are not uncommon in modern devices. We have already anticipated this issue while discussing conditioning circuits in section 1.3.3.

2.3.5. *Stability and frequency response*

In our context, we call a circuit stable when the output to a bounded input is bounded. Other different definitions of stability exist. This definition is usually denoted by the acronym BIBO, from Bounded In Bounded Out. It is often mandatory that a circuit remains stable. In low-power applications, such as operational amplifier circuits, the lack of stability will show up with nonlinearity, saturations, parasitic oscillations and head-scratching problems. In high-power applications, lack of stability may yield expensive repairs, safety hazards, fires, explosions and nuclear meltdowns... If a circuit is unstable, most of the time it is practically useless[2]. In modern voltage-feedback operational amplifiers, there is an internal compensation network that tries to sort out a trade-off between overall speed and stability. The need for stability makes sure that the small signal bandwidth of the operational amplifier is often limited by introducing a low-frequency dominant pole in the differential gain A_d. In most cases, the quest for stability requires the small signal bandwidth of the operational amplifier to be artificially limited. The most widespread technique consists of introducing a low-frequency dominant pole in the open loop differential gain A_d. When a feedback network is tailored so that the closed-loop gain of the circuit is G, then the same circuit has a $-3\,\mathrm{dB}$ bandwidth equal to f_p. The gain-bandwith product $G \times f_\mathrm{p}$ is approximately equal to a constant f_BW, an important characteristic always specified in datasheets of operational amplifiers. In other words, increasing the gain by acting on the feedback around the same operational amplifier entails a reduction of the frequency band treatable by the circuit. There is, however, a notable case in which the

2 With the notable exception of oscillators, where a certain degree of instability is sought and kept under close control in order to initiate and sustain the oscillation.

designer needs to take special care: most operational amplifiers do not appreciate high capacitive loads at their inputs and output (capacitors, long cables, etc.).

If the limitation of the bandwidth is an effect associated with the small signal behavior of the circuits, a large signal (nonlinear) effect is also evident: there is a limitation to the slope of the variation of the output voltage versus the time (the so-called *slew rate*).

TEXAS INSTRUMENTS **LMP7721**

www.ti.com SNOSAW6D – JANUARY 2008 – REVISED MARCH 2013

2.5V Electrical Characteristics (continued)

Unless otherwise specified, all limits are specified for $T_A = 25°C$, $V^+ = 2.5V$, $V^- = 0V$, $V_{CM} = (V^+ + V^-)/2$. **Boldface** limits apply at the temperature extremes.

Symbol	Parameter	Conditions		Min (1)	Typ (2)	Max (1)	Units
I_{BIAS}	Input Bias Current	$V_{CM} = 1V$ (4) (5)	25°C		±3	±20	fA
			−40°C to 85°C			±900	fA
			−40°C to 125°C			**±5**	pA
I_{OS}	Input Offset Current	$V_{CM} = 1V$ (5)			6	40	fA
CMRR	Common Mode Rejection Ratio	$0V \le V_{CM} \le 1.4V$		83 **80**	100		dB
PSRR	Power Supply Rejection Ratio	$1.8V \le V^+ \le 5.5V$ $V^- = 0V$, $V_{CM} = 0$		84 **80**	92		dB
CMVR	Input Common-Mode Voltage Range	CMRR ≥ 80 dB CMRR ≥ 78 dB		−0.3 **−0.3**		1.5 **1.5**	V
A_{VOL}	Large Signal Voltage Gain	$V_O = 0.15V$ to $2.2V$ $R_L = 2\ k\Omega$ to $V^+/2$		88 **82**	107		dB
		$V_O = 0.15V$ to $2.2V$ $R_L = 10\ k\Omega$ to $V^+/2$		92 **88**	120		dB
V_O	Output Swing High	$R_L = 2\ k\Omega$ to $V^+/2$		70 **77**	25		mV from V^+
		$R_L = 10\ k\Omega$ to $V^+/2$		60 **66**	20		mV from V^+
	Output Swing Low	$R_L = 2\ k\Omega$ to $V^+/2$			30	70 **73**	mV
		$R_L = 10\ k\Omega$ to $V^+/2$			15	60 **62**	mV
I_O	Output Short Circuit Current	Sourcing to V^- $V_{IN} = 200$ mV (6)		36 **30**	46		mA
		Sinking to V^+ $V_{IN} = -200$ mV (6)		7.5 **5.0**	15		mA
I_S	Supply Current				1.1	1.5 **1.75**	mA
SR	Slew Rate	$A_V = +1$, Rising (10% to 90%)			9.3		V/µs
		$A_V = +1$, Falling (90% to 10%)			10.8		V/µs
GBW	Gain Bandwidth Product				15		MHz
e_n	Input-Referred Voltage Noise	f = 400 Hz			8		$nV/\sqrt{Hz}$
		f = 1 kHz			7		$nV/\sqrt{Hz}$
i_n	Input-Referred Current Noise	f = 1 kHz			0.01		$pA/\sqrt{Hz}$
THD+N	Total Harmonic Distortion + Noise	f = 1 kHz, $A_V = 2$, $R_L = 100\ k\Omega$ $V_O = 0.9\ V_{PP}$			0.003		%
		f = 1 kHz, $A_V = 2$, $R_L = 600\Omega$ $V_O = 0.9\ V_{PP}$			0.003		%

(4) Positive current corresponds to current flowing into the device.
(5) This parameter is specified by design and/or characterization and is not tested in production.
(6) The short circuit test is a momentary open loop test.

Figure 2.3. *Some of the characteristics of the LMP7721 operational amplifier. Courtesy of Texas Instruments*

2.3.6. *Examples*

Figure 2.3 shows an extract of the data sheet of an operational amplifier optimized for low bias current. Note the typical bias current value of 3 fA at a temperature of 25°C, which rises with the temperature (5 pA at 125°C is a good achievement). Note also how the output saturation voltages are clearly specified: this is a rail-to-rail operational amplifier and this stuff matters. Its performances in terms of noise, gain-bandwidth product and slew rate are also quite sound. Figure 2.3 does not show the input offset voltage, specified elsewhere to be typically ±50 μV at 25°C and less than ±480 μV over an extended temperature range. This is a very decent offset performance, even if is clear that the device is not optimized toward this direction. As an exercise, compare this device with those in the following list (search for the data sheet by yourself):

– the venerable general purpose bipolar μA741, designed in 1968 but still produced today. Compare it with the 1967 vintage LM101. Why has the μA741 been so successful and the LM 101 almost forgotten?

– the JFET-input TL081;

– the first precision bipolar operational amplifier OP07.

Do not forget to search those amplifiers in the online catalog of your favorite electronics dealer. Compare their costs.

2.4. Instrumentation amplifiers

2.4.1. *Introduction*

In Chapter 1, we saw how the output of a sensor conditioning circuit is a voltage, which most of the time needs to be amplified. Moreover, in some situations, the voltage signal carrying information is not single ended (i.e., referred to the reference node) but differential. A classical example is the Wheatstone bridge with resistive sensors: the output signal is a voltage difference between two nodes, as shown in Figure 1.25.

We thus need a circuit able to extract the differential voltage signal without perturbing it. We would also like to be able to easily adjust the differential gain of the circuit in a reasonable range by changing only one component of the circuit. The ability to extract the differential voltage regardless of the common mode voltage is quantified by the common mode rejection ratio parameter, which is ideally very high. In the following sections, we will describe a selection of classical circuits, all based on operational amplifiers.

2.4.2. *Differential amplifier with one operational amplifier*

Figure 2.4 shows the classical differential amplifier made with one operational amplifier. The idea is to apply some amount of feedback in order to tame the differential gain of the operational amplifier (as you remember from section 2.2.1: it is very high, but it is variable, as affected by the frequency, power supply voltage, operating temperature, etc.).

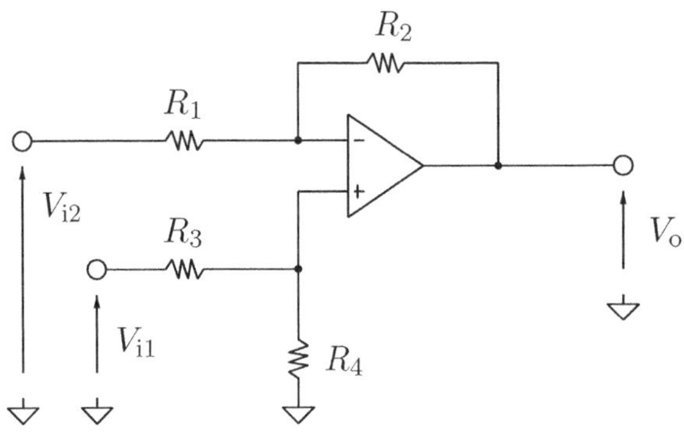

Figure 2.4. *Differential amplifier built using one operational amplifier*

A little bit of circuit analysis, applying the rules given in section 2.2.2, allows us to write down the relationship between voltages at the inputs V_{i1} and V_{i2} and the output V_o:

$$V_o = \frac{R_1 + R_2}{R_1} \times \frac{R_4}{R_3 + R_4} V_{i1} - \frac{R_2}{R_1} V_{i2}. \qquad [2.5]$$

To understand how this circuit can be exploited as a differential amplifier, we rewrite this expression by representing the electrical state of the inputs using the differential and common mode voltages. Thus, we apply the following relations, which can be seen as some sort of a coordinates change:

$$\begin{cases} V_{\rm d} = V_{\rm i1} - V_{\rm i2} \\ V_{\rm cm} = \frac{V_{\rm i1}+V_{\rm i2}}{2} \end{cases} \qquad [2.6]$$

where $V_{\rm d}$ is the differential mode (the signal carrying the information to be extracted) and $V_{\rm cm}$ is the common mode. By inverting the relations, we obtain:

$$\begin{cases} V_{\rm i1} = V_{\rm cm} + \frac{V_{\rm d}}{2} \\ V_{\rm i2} = V_{\rm cm} - \frac{V_{\rm d}}{2} \end{cases} . \qquad [2.7]$$

This yields expressions of $V_{\rm i1}$ and $V_{\rm i2}$ to be inserted in equation [2.5] to obtain equation [2.8]. It relates the output voltage (single ended) to the input differential and common modes of the voltages:

$$\begin{aligned} V_{\rm o} &= \frac{R_1R_4 - R_2R_3}{R_1(R_3+R_4)}V_{\rm cm} \\ &\quad + \frac{R_1+R_2}{2R_1}\left(\frac{R_4}{R_3+R_4} + \frac{R_2}{R_1+R_2}\right)V_{\rm d}. \end{aligned} \qquad [2.8]$$

In the expression [2.8], we recognize the contribution of the differential gain as well as the common mode gain:

$$A_{\rm cm} = \frac{R_1R_4 - R_2R_3}{R_1(R_3+R_4)}, \qquad [2.9]$$

$$A_{\rm d} = \frac{R_1+R_2}{2R_1}\left(\frac{R_4}{R_3+R_4} + \frac{R_2}{R_1+R_2}\right). \qquad [2.10]$$

If a perfect differential amplifier has to be built, resistances $R_1 \ldots R_4$ should be chosen in such a way that the common mode gain is equal to zero. This can be achieved by nulling the numerator of the expression [2.9], thus giving:

$$\frac{R_4}{R_3} = \frac{R_2}{R_1} \qquad [2.11]$$

leading to a simple expression for the differential gain:

$$A_{\mathrm{d}} = \frac{R_2}{R_1}. \qquad [2.12]$$

In practice, very often $R_1 = R_3$ and $R_2 = R_4$, yet perfectly achieving this condition is not possible, because of the inevitable tolerance of the resistances (this takes into account the effect of ageing and thermal drift). In practice, we know that every resistance has a certain relative error from its nominal value. We suppose that the following conditions are verified (worst-case scenario):

– the relative shift of the value of each resistance is equal to the tolerance r;

– the shifts are distributed in such a way that the common mode gain A_{cm} is maximized:

$$\begin{cases} R_1 = R_{1\mathrm{n}}(1+r) \\ R_3 = R_{1\mathrm{n}}(1-r) \\ R_2 = R_{2\mathrm{n}}(1-r) \\ R_4 = R_{2\mathrm{n}}(1+r) \end{cases} \qquad [2.13]$$

where $R_{1\mathrm{n}}$ and $R_{2\mathrm{n}}$ are the nominal values matching condition equation [2.11].

We obtain that the common mode gain is not zero, and it is proportional to the tolerance r:

$$A_{\mathrm{cm}} = \frac{4rR_{2\mathrm{n}}}{R_{1\mathrm{n}} + R_{2\mathrm{n}}}. \qquad [2.14]$$

To calculate the common mode rejection ratio, we suppose that the differential gain has not changed very much if r is small, yielding:

$$C_{\mathrm{mrr}} = 20\log_{10}\frac{A_{\mathrm{d}}}{A_{\mathrm{mc}}} \approx 20\log_{10}\frac{R_{1\mathrm{n}} + R_{2\mathrm{n}}}{4rR_{1\mathrm{n}}}. \qquad [2.15]$$

In the worst-case scenario, this means that by adopting $r = 0.1\%$ tolerance for the resistances and by choosing a gain $A_{\mathrm{d}} = 100$, we might expect that the common mode rejection ratio is about $88\,\mathrm{dB}$.

This circuit has some defects:

– the input impedances are proportional to the values of the resistances. Very high resistance values are, however, associated with noise and the resulting low currents might be sensitive to stray capacitances and couplings;

– the gain can be modified, but the relation [2.11] should always be respected. At least two matched resistances should be varied at the same time to vary the differential gain without disrupting the differential behavior.

To solve the first problem, a second operational amplifier can be added, as discussed in the next section. However, monolithically integrated versions of the circuit are available such as the AD8205. They are very useful for dealing with relatively high common mode voltages that may exceed the supply rails.

2.4.3. *Differential amplifier with two operational amplifiers*

A useful way to vastly increase input impedances is to exploit the excellent input characteristics of operational amplifiers. The circuit shown in Figure 2.5 solves the first issue seen at the end of section 2.4.2, namely the low input impedances. In the circuit, the two inputs are directly connected to the inputs of the operational amplifiers. For this reason, once the correct biasing of the operational amplifiers is assured, inputs are extremely high impedance.

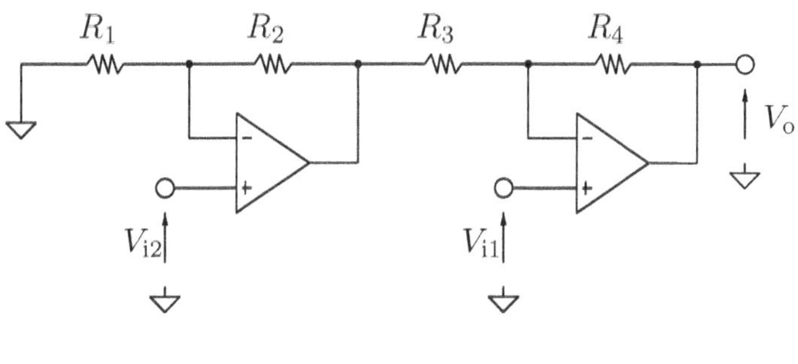

Figure 2.5. *Differential amplifier built using two operational amplifiers*

Analyzing the circuit using the method described in section 2.4.2, we calculate the differential gain, as well as the common mode gain:

$$\begin{cases} A_{\mathrm{d}} = \frac{1}{2}\left[1 + \frac{R_4}{R_3}\left(2 + \frac{R_2}{R_1}\right)\right] \\ A_{\mathrm{cm}} = \left[\frac{R_4+R_3}{R_3} - \frac{R_4}{R_3}\left(1 + \frac{R_2}{R_1}\right)\right]. \end{cases} \qquad [2.16]$$

Nulling the latter, we obtain the balance condition of resistances $R_1 \ldots R_4$ to be respected:

$$\frac{R_1}{R_2} = \frac{R_4}{R_3} \qquad [2.17]$$

thus yielding a simplified expression for the differential gain when the amplifier is purely differential:

$$A_{\mathrm{d}} = 1 + \frac{R_1}{R_2}. \qquad [2.18]$$

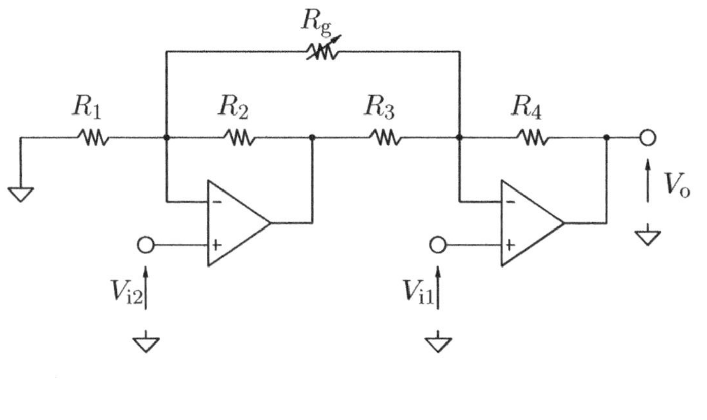

Figure 2.6. *Differential amplifier with variable gain*

At a first sight, it might seem that this circuit is unable to solve the second problem described in the previous section, i.e. the fact that it might not be easy to change the gain by modifying two matched resistances at the same time. In reality, a solution exists connecting an adjustable fifth resistor R_{g}, which allows us to trim the gain without bothering with two matched devices, as shown in Figure 2.6. In this case, the ratio described by equation [2.17] must be respected, but the differential gain can be written as follows:

$$A_{\mathrm{d}} = 1 + 2\frac{R_1}{R_{\mathrm{g}}} + \frac{R_1}{R_2} \quad [2.19]$$

If the two issues of the differential amplifier discussed in section 2.4.2 have been successfully addressed, this circuit still has a more subtle flaw. In fact, the signal paths are not symmetrical: the signal entering from V_{i1} passes through two operational amplifiers, whereas the signal entering from V_{i1} passes through one. When the limitations of the operational amplifiers begin to play an important role (for example when the frequency is relatively high), the asymmetry decreases performances and in particular the common mode rejection ratio of the circuit.

2.4.4. *Differential amplifier with three operational amplifiers*

The circuit shown in Figure 2.7 is a more complex differential amplifier. It is quite commonly used in measuring systems and for this reason, when people say "instrumentation amplifier", they are often referring to this particular circuit. To understand its behavior, we split it into two subcircuits:

– an input stage, which has a differential input and a differential output, meant to boost the differential mode, while leaving the common mode untouched;

– a differential amplifier to provide a single ended output related to the input differential mode.

The second stage is in fact the circuit discussed in section 2.4.2, so we now analyze the input stage as shown in Figure 2.8, which is perfectly symmetrical if $R_1 = R_1'$. If we suppose that the operational amplifiers are ideal, rule 2, seen in section 2.2.2, states that the voltages at the nodes A and B are equal, respectively, to V_{i2} and V_{i1}. By supposing for a moment that V_{o1} and V_{o2} are known, we apply the Millman theorem:

$$\begin{cases} \text{node A: } \dfrac{\frac{V_{o2}}{R_1}+\frac{V_{i1}}{R_g}}{\frac{1}{R_1}+\frac{1}{R_g}} = V_{i2} \\ \\ \text{node B: } \dfrac{\frac{V_{i2}}{R_g}+\frac{V_{o1}}{R'_1}}{\frac{1}{R_g}+\frac{1}{R'_1}} = V_{i1} \end{cases} \qquad [2.20]$$

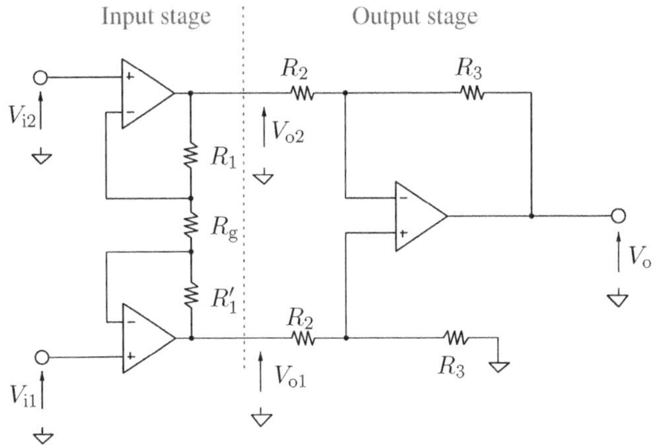

Figure 2.7. *Differential amplifier with three operational amplifier: the instrumentation amplifier by antonomasia*

By rearranging terms, we get:

$$\begin{cases} V_{o1} = \frac{R_g+R'_1}{R_g}V_{i1} - \frac{R'_1}{R_g}V_{i2} \\ \\ V_{o2} = \frac{R_g+R_1}{R_g}V_{i2} - \frac{R_1}{R_g}V_{i1} \end{cases} . \qquad [2.21]$$

Similarly to what is done for the input signals with equations [2.6] and [2.7], we define differential and common mode voltages for the two outputs V_{o1} and V_{o2}:

$$\begin{cases} V'_d = V_{o1} - V_{o2} \\ \\ V'_{cm} = \frac{V_{o1}+V_{o2}}{2} \end{cases} . \qquad [2.22]$$

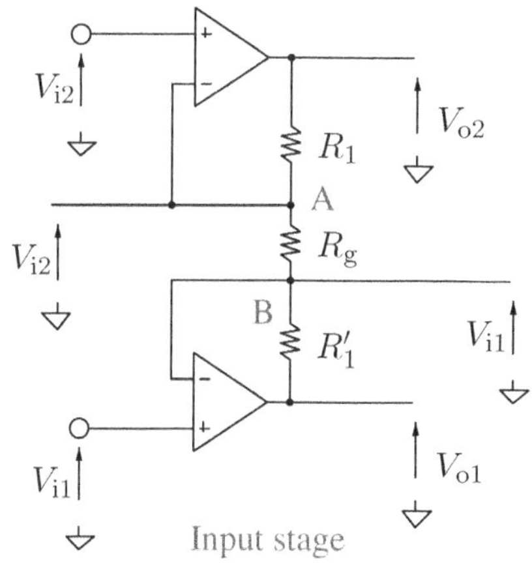

Figure 2.8. *The symmetrical input stage of the instrumentation amplifier shown in Figure 2.7*

However, inevitable tolerances make sure that R_1 and R'_1 are not identical. We consider the worst-case scenario, as follows:

$$\begin{cases} R_1 = R_{1n}(1+r) \\ R'_1 = R_{1n}(1-r) \end{cases} \quad [2.23]$$

where r is the tolerance of the resistances, whose nominal value is R_{1n}. In those conditions, we relate the output common mode with the input common and differential modes. After some algebra, we obtain:

$$V'_{cm} = V_{cm} + \frac{R_{1n}}{R_g} r V_d \quad [2.24]$$

as well as the output differential mode:

$$V'_d = \left(1 + \frac{2R_{1n}}{R_g}\right) V_d. \quad [2.25]$$

We note that:

– if R_g is much smaller than R_{1n}, the input differential mode is greatly amplified;

– the input common mode is mostly not amplified nor attenuated by the first part of the circuit. This first term is very often the most relevant one;

– the output common mode is also affected by the input differential voltage, in a factor which is dependent on the R_{1n}/R_g ratio (the same affecting the differential gain) as well as the tolerance r of the matching between R_1 and R'_1. This second term is often negligible.

To summarize, the first circuit has differential and common mode gains A'_d and A'_{cm} as follows:

$$\begin{cases} A'_d = 1 + \frac{2R_{1n}}{R_g} \\ A'_{cm} \approx 1 \end{cases} . \qquad [2.26]$$

Section 2.4.2 presented the analysis of the second half of the circuit in Figure 2.7:

$$\begin{cases} A''_{mc} = \frac{4rR_{3n}}{R_{2n}+R_{3n}} \\ A''_d = \frac{R_{3n}}{R_{2n}} \end{cases} , \qquad [2.27]$$

where, as usual, r represents the tolerance of the resistances and the "n" subscript indicates their nominal values.

Putting together all of these equations (and neglecting some cross terms) yields the differential and common mode gain of the complete amplifier:

$$\begin{cases} A_{mc} = A'_{mc}A''_{mc} \approx \frac{4rR_{3n}}{R_{2n}+R_{3n}} \\ A_d = A'_dA''_d = \frac{R_{3n}}{R_{2n}}\left(1 + \frac{2R_{1n}}{R_g}\right) \end{cases} . \qquad [2.28]$$

Futhermore, those equations might be simplified when $R_{3n} = R_{2n}$, which is a frequent choice:

$$\begin{cases} A_{mc} \approx 2r \\ A_d = A'_dA''_d = 1 + \frac{2R_{1n}}{R_g} \end{cases} . \qquad [2.29]$$

In fact, the instrumentation amplifier built around three operational amplifiers is both flexible and very convenient to integrate (for example, the INA101, the AD623 and the INA333, among others). In fact, in microelectronics it is difficult to precisely control the absolute value of a passive device, but symmetries such as those required in this circuit can be achieved quite easily. In fact, the end user just needs to choose the gain via the resistance R_g, which is normally to be connected outside of the integrated circuit. This provides outstanding performance, ease of use, as well as flexibility. For example, have a look in Figure 2.9, where the AD623 is described. Compare the expression given for the gain with equation [2.29].

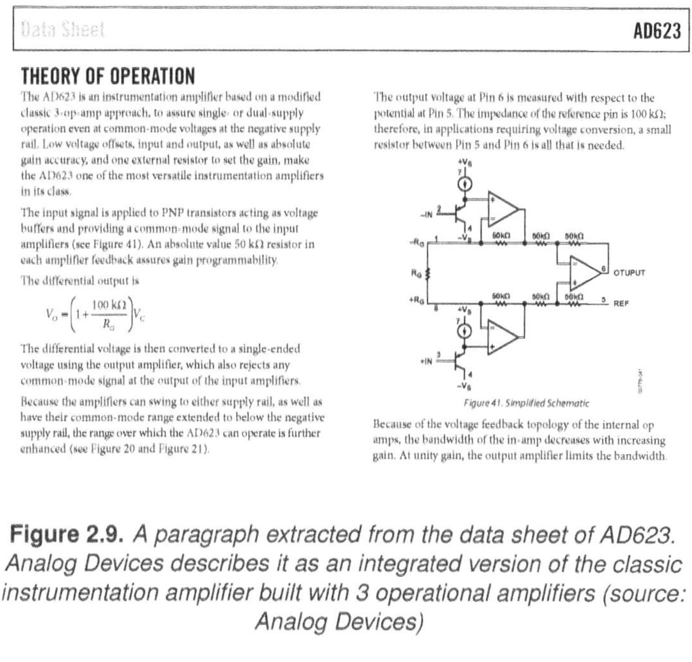
Data Sheet AD623

THEORY OF OPERATION

The AD623 is an instrumentation amplifier based on a modified classic 3-op-amp approach, to assure single- or dual-supply operation even at common-mode voltages at the negative supply rail. Low voltage offsets, input and output, as well as absolute gain accuracy, and one external resistor to set the gain, make the AD623 one of the most versatile instrumentation amplifiers in its class.

The input signal is applied to PNP transistors acting as voltage buffers and providing a common-mode signal to the input amplifiers (see Figure 41). An absolute value 50 kΩ resistor in each amplifier feedback assures gain programmability.

The differential output is

$$V_O = \left(1 + \frac{100\ \text{k}\Omega}{R_G}\right) V_C$$

The differential voltage is then converted to a single-ended voltage using the output amplifier, which also rejects any common-mode signal at the output of the input amplifiers.

Because the amplifiers can swing to either supply rail, as well as have their common-mode range extended to below the negative supply rail, the range over which the AD623 can operate is further enhanced (see Figure 20 and Figure 21).

The output voltage at Pin 6 is measured with respect to the potential at Pin 5. The impedance of the reference pin is 100 kΩ; therefore, in applications requiring voltage conversion, a small resistor between Pin 5 and Pin 6 is all that is needed.

Figure 41. Simplified Schematic

Because of the voltage feedback topology of the internal op amps, the bandwidth of the in-amp decreases with increasing gain. At unity gain, the output amplifier limits the bandwidth.

Figure 2.9. *A paragraph extracted from the data sheet of AD623. Analog Devices describes it as an integrated version of the classic instrumentation amplifier built with 3 operational amplifiers (source: Analog Devices)*

2.5. Isolation amplifiers

Isolation amplifiers are able to effectively decouple two parts of a circuit that must exchange a signal, without having a direct galvanic connection between them. Figure 2.10 shows a schematic view of the way they are done: the signal e_i is transferred through an insulation barrier. Usually, the transfer is done by an optical link (optocouplers), magnetically (transformers) or capacitively. Isolation amplifiers can be required for safety reasons and for protection of equipment, such as when high voltages are involved. In this case, they are effective at eliminating very high common mode voltages V_m between the decoupled sections. This is made explicit in Figure 2.10 by the choice of two different symbols for the reference nodes, once the isolation barrier is crossed. A second important reason is to avoid ground loops, yielding severe electromagnetic compatibility issues (see section 5.8).

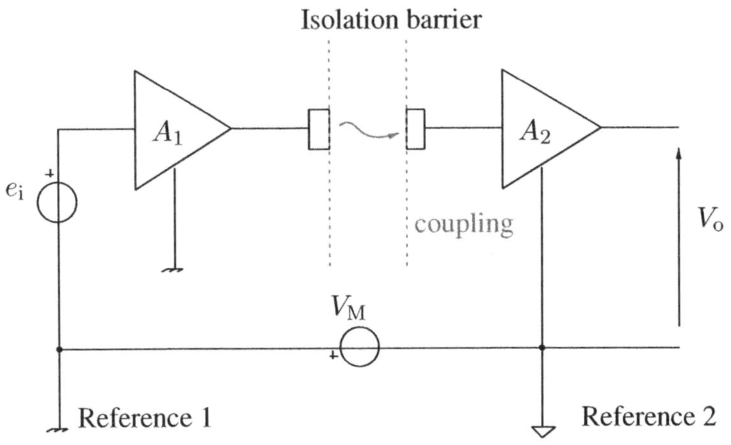

Figure 2.10. *A schematic view of the principles of an isolation amplifier. The presence of an isolation barrier makes sure that the two reference nodes can be subjected to a voltage V_M without any current flowing and with no risk for the signal integrity as long as V_M remains below a certain limit, specified in the data sheet*

Isolation amplifiers might be rated to guarantee several thousand volts of isolation. Examples include the classic ISO120 integrated isolation amplifier, whose internal structure is shown in Figure 2.11. The isolation barrier is capacitive, and the input signal is used to modulate a carrier around 400 kHz. Note how a feedback loop is used on one side of the isolation barrier to achieve good linearity: this trick is effective when it is possible to obtain almost identical circuits on the two sides of the isolation barrier. Figure 2.12 contains the description of the working principle, and the effect of sampling is visible in the oscillograms.

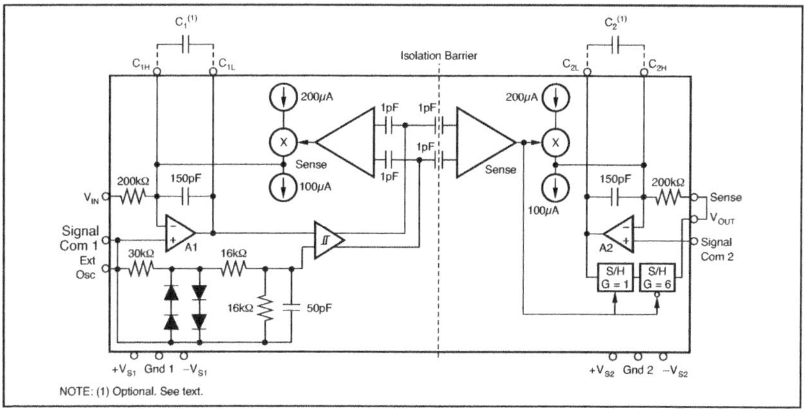

Figure 2.11. *Block diagram of the internal structure of ISO120, a classic isolation amplifier. Courtesy of Texas Instruments*

An example of a low-cost modern device proposed by Texas Instruments is the AMC1100, once again with a capacitive coupling. On the other side, the Analog Devices AD202 features a complete amplifier module, transformer-coupled, with an onboard-isolated power supply converter and a cost, which is aligned with the performance.

2.6. Conclusion

In this chapter, we rapidly covered the main characteristics of operational amplifiers. We then discussed the desirable characteristics of instrumentation amplifiers and described three variants, often employed in practical situations. We concluded our discussion by presenting isolation amplifiers.

TYPICAL PERFORMANCE CURVES (CONT)

$T_A = +25°C$; $V_{S1} = V_{S2} = ±15V$; and $R_L = 2kΩ$, unless otherwise noted.

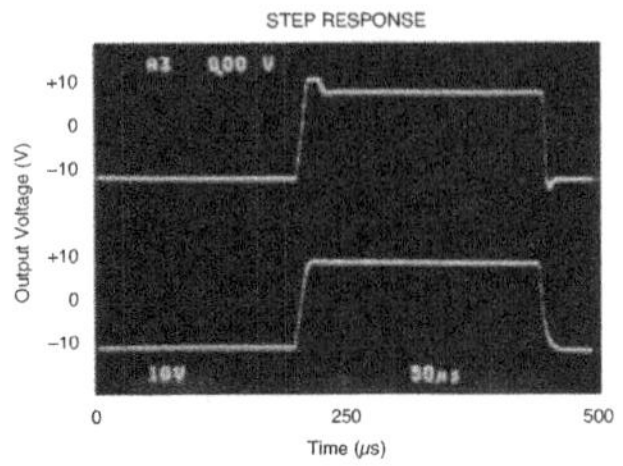

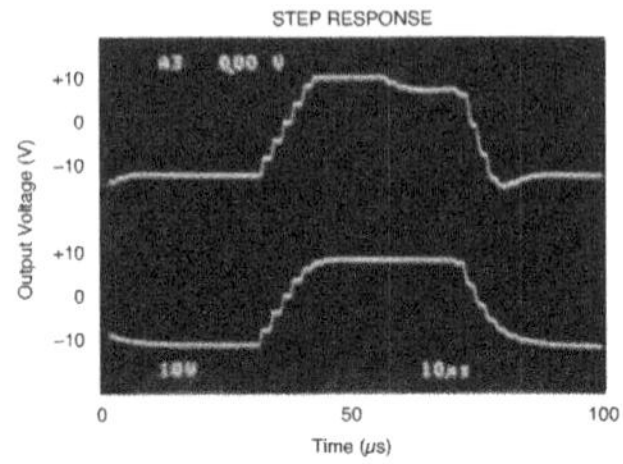

THEORY OF OPERATION

The ISO120 and ISO121 isolation amplifiers comprise input and output sections galvanically isolated by matched 1pF capacitors built into the ceramic barrier. The input is duty-cycle modulated and transmitted digitally across the barrier. The output section receives the modulated signal, converts it back to an analog voltage and removes the ripple component inherent in the demodulation. The input and output sections are laser-trimmed for exceptional matching of circuitry common to both input and output sections.

FREE-RUNNING MODE

An input amplifier (A1, Figure1) integrates the difference between the input current ($V_{IN}/200kΩ$) and a switched ±100μA current source. This current source is implemented by a switchable 200μA source and a fixed 100μA current sink. To understand the basic operation of the input section, assume that $V_{IN} = 0$. The integrator will ramp in one direction until the comparator threshold is exceeded. The comparator and sense amp will force the current source to switch; the resultant signal is a triangular waveform with a 50% duty cycle. If V_{IN} changes, the duty cycle of the integrator will change to keep the average DC value at the output of A1 near zero volts. This action converts the input voltage to a duty-cycle modulated triangular waveform at the output of A1 near zero volts. This action converts the input voltage to a duty-cycle modulated triangular waveform at the output of A1 with a frequency determined by the internal 150pF capacitor. The comparator generates a fast rise time square wave that is simultaneously fed back to keep A1 in charge balance and also across the barrier to a differential sense amplifier with high common-mode rejection characteristics. The sense amplifier drives a switched current source surrounding A2. The output stage balances the duty-cycle modulated current against the feedback current through the 200kΩ feedback resistor, resulting in an average value at the Sense pin equal to V_{IN}. The sample and hold amplifiers in the output feedback loop serve to remove undesired ripple voltages inherent in the demodulation process.

SYNCHRONIZED MODE

A unique feature of the ISO120 and ISO121 is the ability to synchronize the modulator to an external signal source. This capability is useful in eliminating trouble-some beat frequencies in multi-channel systems and in rejecting AC signals and their harmonics. To use this feature, external capacitors are connected at C_1 and C_2 (Figure 1) to change the free-running carrier frequency. An external signal is applied to the Ext Osc pin. This signal forces the current source to switch at the frequency of the external signal. If V_{IN} is zero, and the external source has a 50% duty cycle, operation proceeds as described above, except that the switching frequency is that of the external source. If the external signal has a duty cycle other than 50%, its average value is not zero. At start-up, the current source does not switch until the integrator establishes an output equal to the average DC value of the external signal. At this point, the external signal is able to trigger the current source, producing a triangular waveform, symmetrical about the new DC value, at the output of A1. For $V_{IN} = 0$, this waveform has a 50% duty cycle. As V_{IN} varies, the waveform retains its DC offset, but varies in duty cycle to maintain charge balance around A1. Operation of the demodulator is the same as outlined above.

Synchronizing to a Sine or Triangle Wave External Clock

The ideal external clock signal for the ISO120/121 is a ±4V sine wave or ±4V, 50% duty-cycle triangle wave. The *ext osc* pin of the ISO120/121 can be driven directly with a ±3V to ±5V sine or 25% to 75% duty-cycle triangle wave and the ISO amp's internal modulator/demodulator circuitry will synchronize to the signal.

Synchronizing to signals below 400kHz requires the addition of two external capacitors to the ISO120/121. Connect one capacitor in parallel with the internal modulator capacitor and connect the other capacitor in parallel with the internal demodulator capacitor as shown in Figure 1.

Figure 2.12. *Another extract of ISO120 data sheet. Here is Ti's description of how the device works. Courtesy of Texas Instruments*

3

Elements of Active Filter Synthesis

3.1. Introduction

One of the most convenient operations in signals is linear filtering. In fact, useful information is conveyed on a certain spectral band and everything outside that band can be considered noise. Filtering it out is a good way to increase overall measurement quality. Furthermore, modern measurement systems often involve a sample and hold circuit, coupled with an analog to digital converter. As sampling is subjected to the aliasing phenomenon, this means that an analog anti-aliasing filter *must always be present.* In some low-performance applications, the natural low-pass behavior of sensors and amplifiers might be sufficient, but to achieve high performances, a dedicated filtering block becomes mandatory.

Filter design is a huge domain whose surface is barely scratched by this chapter. We want to provide an understanding of the problems and the terminology, as well as a design method for simple active filters.

Figure 3.1 shows the classic design flow for filter design. The first operation is called approximation. It consists of writing down a transfer function $H(p)$ of a two-port network (in the Laplace domain):

$$H(p) = \frac{n(p)}{g(p)}, \qquad [3.1]$$

where $n(p)$ and $g(p)$ are polynomials (with real coefficients of p, the Laplace variable), such that the response $|H(\mathrm{j}2\pi f)|$ of the filter matches some constraints. Usually, attenuation in some frequency bands is requested, whereas other spectral components must be left untouched. Alternatively, constraints may involve the pulse response, for which damped oscillations may be allowed or not, or may involve achieving a group delay that is as constant as possible.

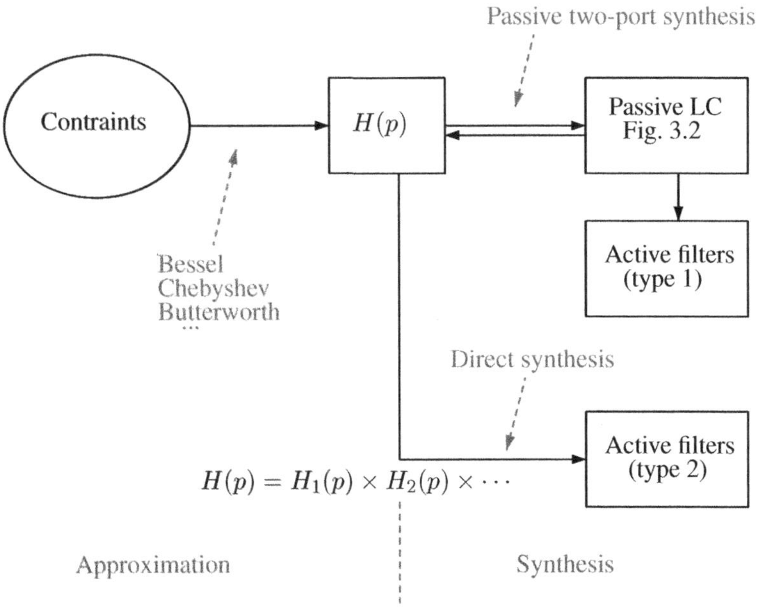

Figure 3.1. *Classical design flow of analog filters*

Once the transfer function is written, a first possibility is to synthesize a passive two-port network (usually composed only of capacitors and inductors), obtaining a circuit similar to the one shown in Figure 3.2. This is the most classical filter approach and there is a difficulty here because, in general, an arbitrary transfer function cannot always be synthesized using only passive lumped devices. Moreover, each component affects the overall filter response. Therefore, in the synthesis itself, one must deal with practical constraints for writing down an acceptable $H(p)$. Many different cases have been tackled successfully and very complete catalogs exist, such as the classic Zverev book. This first class of filters will not be treated here (have a look at the tables of [ZVE 67] for more details). Those filters usually show interesting properties such as minimum sensitivity to component tolerances (Orchard's theorem [ORC 66]) and interesting noise properties.

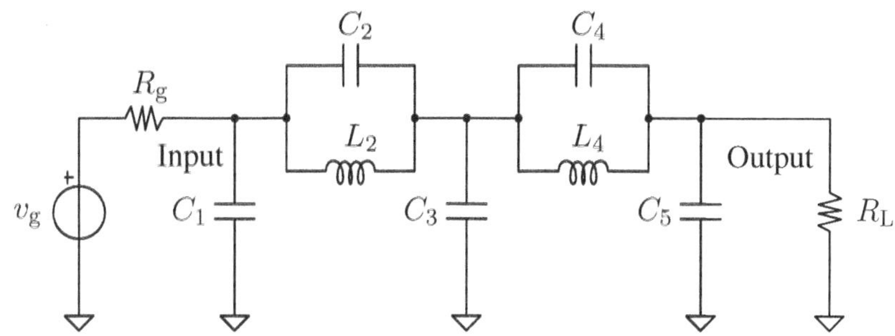

Figure 3.2. *Example of an LC passive filter: each capacitor and inductor affects the overall behavior of the filter, which makes its design quite complex. Moreover, the filter must be calculated for precise values of source and load impedances, respectively, R_g and R_L*

The use of inductors is often inconvenient, especially for low-frequency circuits, and therefore a first class of active filters (arbitrarily called "type 1" in Figure 3.1) is obtained by emulating passive inductors via active circuits such as gyrators or general impedance converters (also known as GICs).

The second possible strategy, which will be described later in this chapter, relies on the direct synthesis of the transfer function by means of cascaded active cells. They have the remarkable property that they can be fabricated such that cascading two or more cells does not change their individual behavior. The idea is, therefore, to split the transfer function into a product of low-order terms, each one being synthesized separately by a cell.

In this chapter, we present some elements of synthesis strategy as well as some tables of filter coefficients useful for active filters ("type 2" in Figure 3.1). We focus on all-pole low-pass filters, and we briefly see how other kinds of filters can be traced back to an equivalent low-pass filter calculation.

3.2. Low-pass filter approximation

3.2.1. *Aliasing in sampled systems and anti-aliasing filters*

In the vast majority of cases, a measurement system will involve analog to digital converters, which sample an analog voltage resulting from the measurement. Often[1], this is done at a constant rate with a sampling frequency F_E. The well-known Nyquist–Shannon sampling theorem states that the maximum frequency that can be reconstructed in this case from the obtained samples is equal to $F_E/2$. Without giving all the mathematical details, which can be found in signal processing textbooks, one can intuitively expect that some interesting situations may arise when one tries to sample a signal whose frequency is higher than $F_E/2$. In fact, a phenomenon called *aliasing* manifests itself when the Nyquist–Shannon theorem is not respected and some spurious signals appear: their frequencies are translated inside the band comprised between 0 and $F_E/2$ and may superimpose with the useful signal, entailing a loss of information.

One example can help visualize the effect of aliasing: you have certainly seen a car accelerating watching a movie or a YouTube video. You may probably have noticed that the car's wheels at first move in the "correct" direction when the car moves slowly, but they may begin to spin the other way round when the car's speed passes a certain threshold. What happens here is that the camera is sampling the image of the real car and the Nyquist–Shannon theorem is not respected anymore after the wheels spin past a certain speed, given the frame rate of the camera (i.e. the sampling frequency F_E). So the aliasing effect on some details (the rims) manifests itself by scrambling the frequency

1 Yet not necessarily!

at which they appear to be moving and you get the "apparent" incorrect backwards movement.

Aliasing is not a bad thing by itself and in some cases it may be employed creatively (this technique is called undersampling). However, to avoid it, signals have to be filtered in such a way that we ensure the Nyquist–Shannon theorem is respected. The filter must be applied before the signals are sampled and therefore an anti-aliasing filter is always an analog low-pass filter.

3.2.2. *Definitions*

Figure 3.3 shows a typical request for a low-pass filter: the gain response should lie in the unshaded area of the diagram.

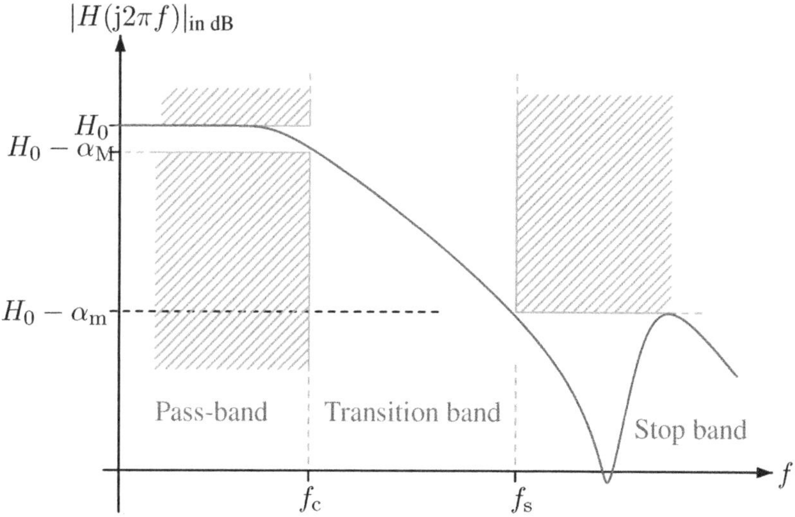

Figure 3.3. *Attenuation behavior requested from a low-pass filter: the filter response should not lie in the shaded area. A possible filter approximation is shown, which has some ripple in the stopband*

For a low-pass filter, we define the following parameters:

– α_{M} is the maximum allowed attenuation (in dB) in the passband. This term, therefore, gives the limit of the so-called attenuation ripple allowed in the passband;

– f_{c} is the maximum frequency of the passband (cutoff frequency);

– f_{s} is the frequency at which the stopband begins;

– α_{m} is the minimum guaranteed attenuation (in dB) in the stopband.

The choice of those parameters (as well as the type of approximation, described below) constitutes an important and difficult step in the filter design, especially when considerable freedom is possible. In fact, it appears that writing the transfer function [3.1] is an indeterminate problem and therefore is remarkably difficult. In fact, how can we choose the degree of the numerator $n(p)$ and denominator $g(p)$ polynomials? If higher degrees imply higher complexity in circuit realization, how can we obtain acceptable performances with a circuit with a reasonable number of lumped components? How can we choose the coefficients of the polynomials yielding a circuit which can be synthesized? Therefore, filter approximation and synthesis are more like a constrained optimization problem than a procedure to be followed blindly.

3.2.3. *All-pole filters: normalization and factorization*

First of all, we restrict our analysis to all-pole filters, i.e. those having a transfer function, which can be written as follows:

$$H(p) = \frac{M}{g(p)}, \qquad [3.2]$$

where $g(p)$ is a polynomial of degree n and M is a constant. Polynomial coefficients must be real numbers (if not, we cannot expect to build a circuit having $H(p)$ as a transfer function) and $g(p)$ must have the following property:

$$\max_{\omega}\left(\frac{1}{|g(\mathrm{j}\omega)|}\right) = 1. \qquad [3.3]$$

Furthermore, we normalize all frequencies by dividing them by a normalization factor, often related to f_c (except as we will see later for Bessel–Thompson filters, where it is related to the group delay τ_g). We will be more precise about that later. For the moment, it is enough to know that a normalization can be done and to use s and p to indicate, respectively, the Laplace normalized and non-normalized variable. Of course $g(p)$ being a polynomial with real coefficients with a n degree, we expect it to have exactly $n \in \mathbb{N}$ solutions, with Γ them being real ($r \in \mathbb{N}$, $r \leq n$). If complex solutions exist, they must come in $c \in \mathbb{N}$ couples (complex-conjugate) and $n = r+2c$. Therefore, the normalized transfer function $H_n(s)$ can be factored as follows:

$$
\begin{aligned}
H_n(s) = \frac{M}{g(s)} &= \frac{M}{\prod_{i\leq r}(s - s_{r,i}) \prod_{j\leq c}(\frac{1}{\omega_{s,j}^2} s^2 + \frac{1}{q_{s,j}\omega_{s,j}} s + 1)} \\
&= M \prod_{i\leq r} K_i(s) \prod_{j\leq c} T_j(s).
\end{aligned}
\qquad [3.4]
$$

Each first-order term (real root) is as follows:

$$K_i(s) = \frac{1}{s - s_{r,i}}. \qquad [3.5]$$

Each second-order (complex conjugate roots, low-pass) term can be written as follows:

$$T_j(s) = \frac{\omega_{s,j}^2}{s^2 + \frac{\omega_{s,j}}{q_{s,j}} s + \omega_{s,j}^2} = \frac{1}{\frac{1}{\omega_{s,j}^2} s^2 + \frac{1}{q_{s,j}\omega_{s,j}} s + 1}. \qquad [3.6]$$

To obtain a real-coefficient polynomial, such poles of $H_n(s)$ should be complex conjugate pairs (we omit the j-index for simplicity): $s_{1,2} = a \pm \mathrm{j}b$. The pole frequency is as follows:

$$\omega_s = \sqrt{a^2 + b^2}, \qquad [3.7]$$

and the quality factor is given by:

$$q_s = -\frac{\omega_s}{2a}. \qquad [3.8]$$

The damping factor $\zeta = 1/(2q_\mathrm{s}) = -a/\omega_\mathrm{s}$ is also often employed.

We note that the approximation problem is completely solved once we know the position of the poles of $H_n(s)$ in the complex plane, i.e. $s_{r,i}$ for $i \leq r$ as well as $\omega_{s,j}$ and $q_{s,j}$ for $j \leq c$.

Three classical approximations are described in this document:

– Butterworth filters, where the optimization seeks for a maximally flat behavior in passband.

– Chebyshev filters, where a certain amount of attenuation ripple is tolerated in the passband, but the attenuation rises sharply outside the passband.

– Bessel–Thompson filters, where the group delay is the best possible approximation for a constant τ in the band-pass.

The characteristics and the coefficients of the different approximations will be discussed in the detail in following sections. However, Figures 3.4 and 3.5 compare their typical sixth-degree normalized response. It is visible how achieving a “sharp” cutoff in the attenuation is clearly not the goal of a Bessel–Thompson optimization, whereas it is comparatively easy to obtain selective filters with Chebyshev's criterion[2].

The situation changes when observing the group delay, where Chebyshev filters show a highly variable behavior, resulting in a certain degree of ringing that may become evident when processing impulsive signals. *Only focusing on attenuation is a common mistake, giving an incomplete picture of the filter behavior.*

2 The slope of the attenuation versus frequency behavior is in reality different only for frequencies close to the cutoff: this is practically relevant, since here is usually where the transition band is. At much higher frequencies, the slope becomes only dependant on the order of the filter.

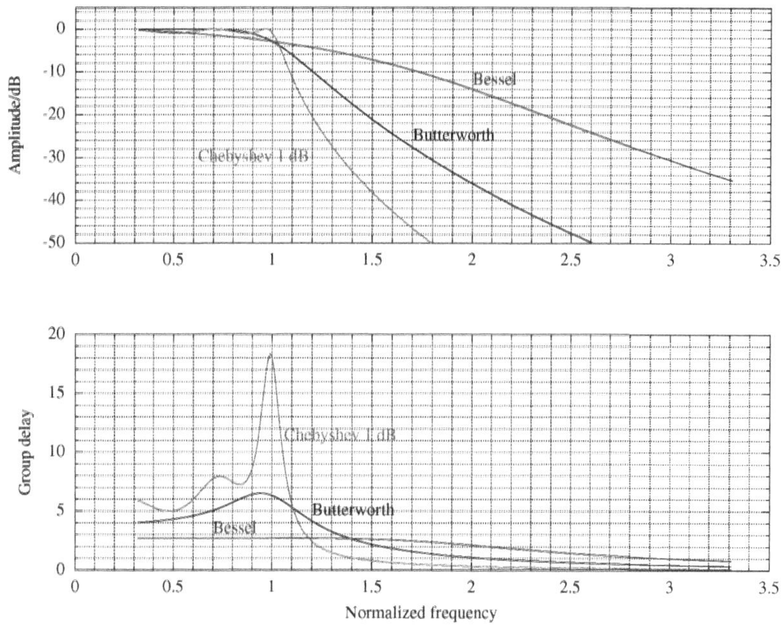

Figure 3.4. *Gain and group-delay responses for sixth-order Butterworth, Chebyshev* 1 dB *and Bessel–Thompson filters. The group delay* τ_g *for Bessel filters has been adjusted such that a* -3 dB *gain is reached at the normalized frequency 1*

3.2.4. *Butterworth approximation*

The transfer function in Butterworth filters is optimized to achieve a maximally flat frequency response in the passband. The goal being the flatness, they tend not to be very selective, i.e. given a certain order, once the signal frequency is greater than the cutoff frequency f_c, the attenuation increases, but not very rapidly. The frequency f_c is almost always defined as the frequency at which the attenuation is equal to 3 dB and for the normalization:

$$s = \frac{p}{2\pi f_c}. \quad [3.9]$$

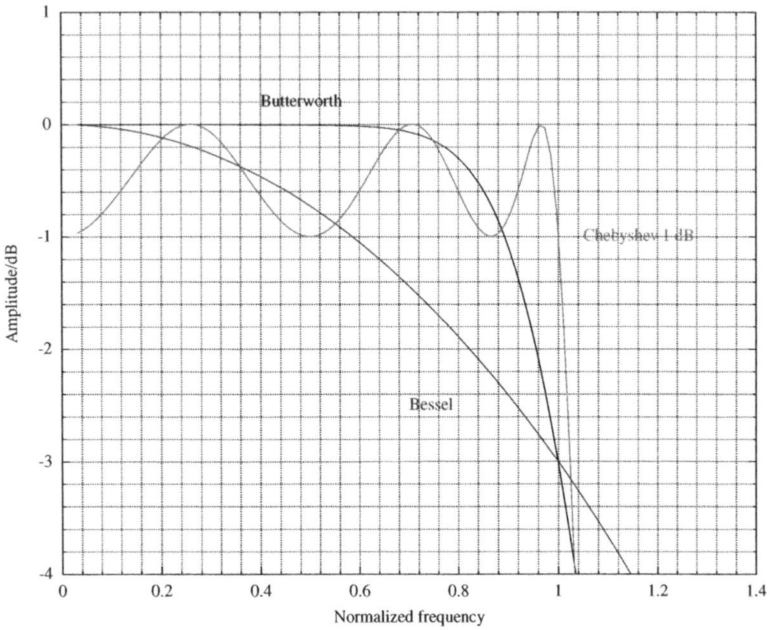

Figure 3.5. *Gain for Butterworth, Chebyshev* $1\,\mathrm{dB}$ *and Bessel–Thompson filters in the band-pass. Zoom of the gain behavior of filters shown in Figure 3.4; for Chebyshev filters, the* $-3\,\mathrm{dB}$ *frequency is not the end of the passband. Application of equation* [3.20] *gives* 1.023442*, a value confirmed graphically*

We define:

$$\epsilon = \sqrt{10^{\alpha_\mathrm{M}/10} - 1}, \qquad [3.10]$$

the minimum order n to satisfy to the guaranteed attenuation in the stopband criterion is as follows:

$$n = \left\lceil \frac{\log \frac{\sqrt{10^{\alpha_\mathrm{m}/10} - 1}}{\epsilon}}{\log(f_\mathrm{s}/f_\mathrm{c})} \right\rceil . \qquad [3.11]$$

The $g(s)$ polynomial can be generated as follows:

$$g(s) = \epsilon \prod_{k=1}^{n} (s - s_k), \qquad [3.12]$$

where the roots s_k are given by the following expression:

$$s_k = \epsilon^{-1/n} \left[-\sin \frac{(2k-1)\pi}{2n} + \mathrm{j} \cos \frac{(2k-1)\pi}{2n} \right]. \qquad [3.13]$$

Table 3.1 shows normalized roots for Butterworth approximation, up to the eighth order.

3.2.5. *Chebyshev approximation*

This type of approximation tries to increase selectivity by tolerating a certain amount of ripple in the band-pass for the module of the transfer function[3]. The tolerated ripple is quantified by α_{M} and different filters can be built for different values of α_{M}. Usually, α_{M} is equal or below $1\,\mathrm{dB}$ and the cutoff frequency f_{c} is defined as the frequency at which the attenuation is equal to α_{M}. So, practically always, f_{c} *is not the frequency corresponding to* $3\,\mathrm{dB}$ *attenuation*, as visible in Figure 3.5. The normalization is usually done as follows:

$$s = \frac{p}{2\pi f_{\mathrm{c}}}. \qquad [3.14]$$

We define:

$$\epsilon = \sqrt{10^{\alpha_{\mathrm{M}}/10} - 1}, \qquad [3.15]$$

3 The goal is approximating a constant attenuation in the bandpass with the so-called Minimax approach. Even more selective filters exist, such as inverse Chebyshev, elliptic (Cauer) filters... They are not all-pole filters: both poles and zeros are present in the transfer function, so this class of filters falls outside the scope of this introductory document (see, for example, [ZVE 67]).

the minimum order n to be adopted for a given constraint in the attenuation is as follows:

$$n = \left\lceil \frac{\operatorname{arcosh} \frac{\sqrt{10^{\alpha_m/10}-1}}{\epsilon}}{\operatorname{arcosh}(f_s/f_c)} \right\rceil . \qquad [3.16]$$

Butterworth, roots s_k for $\alpha_M = 3\,dB$					
3	**4**	**5**	**6**	**7**	**8**
−0.50040 + 0.86671j	−0.38291 + 0.92443j	−0.30916 + 0.95151j	−0.25892 + 0.96631j	−0.22260 + 0.97526j	−0.19515 + 0.98108j
−1.00000	−0.92443 + 0.38291j	−0.80940 + 0.58806j	−0.70739 + 0.70739j	−0.62370 + 0.78210j	−0.55574 + 0.83172j
−0.50040 − 0.86671j	−0.92443 − 0.38291j	−1.00000	−0.96631 + 0.25892j	−0.90127 + 0.43403j	−0.83172 + 0.55574j
	−0.38291 − 0.92443j	−0.80940 − 0.58806j	−0.96631 − 0.25892j	−1.00000	−0.98108 + 0.19515j
		−0.30916 − 0.95151j	−0.70739 − 0.70739j	−0.90127 − 0.43403j	−0.98108 − 0.19515j
			−0.25892 − 0.96631j	−0.62370 − 0.78210j	−0.83172 − 0.55574j
				−0.22260 − 0.97526j	−0.55574 − 0.83172j
					−0.19515 − 0.98108j
Chebyshev, roots s_k for $\alpha_M = 0.1\,dB$ ($\epsilon = 0.15262$):					
3	**4**	**5**	**6**	**7**	**8**
−0.48470 + 1.20616j	−0.26416 + 1.12261j	−0.16653 + 1.08037j	−0.11469 + 1.05652j	−0.08384 + 1.04183j	−0.06398 + 1.03218j
−0.96941	−0.63773 + 0.46500j	−0.43599 + 0.66771j	−0.31335 + 0.77343j	−0.23492 + 0.83549j	−0.18220 + 0.87504j
−0.48470 − 1.20616j	−0.63773 − 0.46500j	−0.53891	−0.42804 + 0.28309j	−0.33947 + 0.46366j	−0.27268 + 0.58468j
	−0.26416 − 1.12261j	−0.43599 − 0.66771j	−0.42804 − 0.28309j	−0.37678	−0.32165 + 0.20531j
		−0.16653 − 1.08037j	−0.31335 − 0.77343j	−0.33947 − 0.46366j	−0.32165 − 0.20531j
			−0.11469 − 1.05652j	−0.23492 − 0.83549j	−0.27268 − 0.58468j
				−0.08384 − 1.04183j	−0.18220 − 0.87504j
					−0.06398 − 1.03218j
Chebyshev, roots s_k for $\alpha_M = 0.5\,dB$ ($\epsilon = 0.34931$):					
3	**4**	**5**	**6**	**7**	**8**
−0.31323 + 1.02193j	−0.17535 + 1.01625j	−0.11196 + 1.01156j	−0.07765 + 1.00846j	−0.05700 + 1.00641j	−0.04362 + 1.00500j
−0.62646	−0.42334 + 0.42095j	−0.29312 + 0.62518j	−0.21214 + 0.73824j	−0.15972 + 0.80708j	−0.12422 + 0.85200j
−0.31323 − 1.02193j	−0.42334 − 0.42095j	−0.36232	−0.28979 + 0.27022j	−0.23080 + 0.44789j	−0.18591 + 0.56929j
	−0.17535 − 1.01625j	−0.29312 − 0.62518j	−0.28979 − 0.27022j	−0.25617	−0.21929 + 0.19991j
		−0.11196 − 1.01156j	−0.21214 − 0.73824j	−0.23080 − 0.44789j	−0.21929 − 0.19991j
			−0.07765 − 1.00846j	−0.15972 − 0.80708j	−0.18591 − 0.56929j
				−0.05700 − 1.00641j	−0.12422 − 0.85200j
					−0.04362 − 1.00500j
Chebyshev, roots s_k for $\alpha_M = 1.0\,dB$ ($\epsilon = 0.50885$):					
3	**4**	**5**	**6**	**7**	**8**
−0.24709 + 0.96600j	−0.13954 + 0.98338j	−0.08946 + 0.99011j	−0.06218 + 0.99341j	−0.04571 + 0.99528j	−0.03501 + 0.99645j
−0.49417	−0.33687 + 0.40733j	−0.23421 + 0.61192j	−0.16988 + 0.72723j	−0.12807 + 0.79816j	−0.09970 + 0.84475j
−0.24709 − 0.96600j	−0.33687 − 0.40733j	−0.28949	−0.23206 + 0.26618j	−0.18507 + 0.44294j	−0.14920 + 0.56444j
	−0.13954 − 0.98338j	−0.23421 − 0.61192j	−0.23206 − 0.26618j	−0.20541	−0.17600 + 0.19821j
		−0.08946 − 0.99011j	−0.16988 − 0.72723j	−0.18507 − 0.44294j	−0.17600 − 0.19821j
			−0.06218 − 0.99341j	−0.12807 − 0.79816j	−0.14920 − 0.56444j
				−0.04571 − 0.99528j	−0.09970 − 0.84475j
					−0.03501 − 0.99645j

Table 3.1. *Normalized roots of $g(s)$ for Butterworth and Chebyshev approximations*

The $g(s)$ polynomial can be generated as follows:

$$g(s) = 2^{n-1}\epsilon \prod_{k=1}^{n}(s - s_k), \quad [3.17]$$

where the roots s_k are given by:

$$s_k = -\sinh A \sin\frac{(2k-1)\pi}{2n} + \mathrm{j}\cosh A \cos\frac{(2k-1)\pi}{2n}, \quad [3.18]$$

where:

$$A = \frac{1}{n}\operatorname{arcsinh}(\epsilon^{-1}). \quad [3.19]$$

If one wants to calculate the $-3\,\mathrm{dB}$ frequency $f_{-3\,\mathrm{dB}}$ for the filter response, the following equation can be used:

$$f_{-3\,\mathrm{dB}} = \cosh\left[\frac{1}{n}\operatorname{arcosh}(\epsilon^{-1})\right]. \quad [3.20]$$

An example of numerical calculations is visible in Figure 3.5, for a particular 1 dB Chebyshev-type response normalized with respect to f_c. Table 3.1 shows normalised roots for Chebyshev approximations, up to the eighth order.

3.2.6. *Bessel–Thompson approximation*

A filter transfer function $H(p)$ is a meromorphic[4] complex function in the Laplace domain. Therefore, a module $|H(p)|$ and a phase $\mathrm{Arg}[H(p)]$ can be determined for each frequency $p = \mathrm{j}\omega = \mathrm{j}2\pi f$. Chebyshev and Butterworth approximations deal with the behavior of the module of the transfer function. Bessel–Thompson approximation is oriented toward the group delay τ_g, closely related to the phase of $H(s)$:

$$\tau_g(\omega) = -\frac{\mathrm{d}\,\mathrm{Arg}[H(\mathrm{j}\omega)]}{\mathrm{d}\omega}. \quad [3.21]$$

4 Meaning it is analytic (differentiable) everywhere in the complex plane, except in a finite and discrete set of points which are poles.

This class of filter tries to achieve the best possible approximation of a constant τ for the group delay $\tau_g(\omega)$ in the passband. Therefore, the normalization is done on the group delay τ:

$$s = \tau p. \tag{3.22}$$

Bessel–Thompson filters are precious when filtering must be done on a signal whose phase must not be distorted, for example in a pulse-code modulation. The polynomial expression is as follows:

$$g(s) = \frac{B_k(s)}{B_k(0)}, \tag{3.23}$$

where B_k represents reverse Bessel polynomials, as shown in Table 3.2.

Degree	Polynomial
0	$B_0(s) = 1$
1	$B_1(s) = s + 1$
2	$B_2(s) = s^2 + 3s + 3$
3	$B_3(s) = s^3 + 6s + 15s + 15$
4	$B_4(s) = s^4 + 10s^3 + 45s^2 + 105s + 105$
5	$B_5(s) = s^5 + 15s^4 + 105s^3 + 420s^2 + 945s + 945$
6	$B_6(s) = s^6 + 21s^5 + 210s^4 + 1260s^3 + 4725s^2 + 10395s + 10395$
Degree	**Roots**
1	-1
2	$-1.500000 \pm j0.866025$
3	-2.322185
	$-1.838907 \pm j1.754381$
4	$-2.896211 \pm j0.867234$
	$-2.103789 \pm j2.657418$
5	-3.646739
	$-3.351956 \pm j1.742661$
	$-2.324674 \pm j3.571023$
6	$-4.248359 \pm j0.867509$
	$-3.735708 \pm j2.626272$
	$-2.515932 \pm j4.492673$

Table 3.2. *Normalized reverse Bessel polynomials and their roots*

3.2.7. *Examples*

3.2.7.1. *Fourth-order* $1\,\mathrm{kHz}$ *Butterworth low-pass approximation*

This section gives an example of a filter approximation (i.e. how to write down the transfer function $H(p)$) using the Butterworth low-pass model, aiming for a cutoff frequency $f_\mathrm{c} = 1\,\mathrm{kHz}$. The first step is to write down the normalized low-pass transfer function. From tables, normalized poles of the transfer function are:

Poles	**Pole frequency**	q**-factor**
$s_{1,2} = -0.38291 \pm \mathrm{j}0.92443$	$\omega_{\mathrm{s},1,2} = 1.0006$	$q_{\mathrm{s},1,2} = 1.3058$
$s_{3,4} = -0.92443 \pm \mathrm{j}0.38291$	$\omega_{\mathrm{s},3,4} = 1.0006$	$q_{\mathrm{s},3,4} = 0.54087$

In our case, $\omega_{\mathrm{s},1,2} = \omega_{\mathrm{s},3,4}$ is 1.0006 and not 1 because of the truncation to five figures of results in the tables. So we might safely use 1 for both pole frequencies in the following calculations. *It is a general property of Butterworth filters that all poles of the normalized transfer function lies in the unitary circle on the complex plane.*

The normalized transfer function is composed of two second-order terms. From equation [3.6], therefore $H_\mathrm{n}(s)$ is as follows:

$$H_\mathrm{n}(s) = \frac{1}{s^2 + \frac{1}{1.3058}s + 1} \times \frac{1}{s^2 + \frac{1}{0.54087}s + 1}. \qquad [3.24]$$

The last thing that remains to be done is to denormalize the transfer function by applying the following transformation:

$$s = \frac{p}{2\pi f_\mathrm{c}} = \frac{p}{2\pi \times 1\,\mathrm{kHz}}, \qquad [3.25]$$

therefore obtaining:

$$H(p) = \frac{1}{\left(\frac{p}{2\pi\times 1\,\mathrm{kHz}}\right)^2 + \frac{1}{1.3058} \times \frac{p}{2\pi\times 1\,\mathrm{kHz}} + 1} \times \frac{1}{\left(\frac{p}{2\pi\times 1\,\mathrm{kHz}}\right)^2 + \frac{1}{0.54087} \times \frac{p}{2\pi\times 1\,\mathrm{kHz}} + 1}, \qquad [3.26]$$

which can be further developed as follows:

$$H(p) = \frac{1}{2.5330 \times 10^{-8}\,\mathrm{s}^2\,p^2 + 1.2188 \times 10^{-4}\,\mathrm{s}\,p + 1} \times \frac{1}{2.5330 \times 10^{-8}\,\mathrm{s}^2\,p^2 + 2.9426 \times 10^{-4}\,\mathrm{s}\,p + 1}. \qquad [3.27]$$

Some may find employing measurement units in equations strange, as we did in equations [3.26] and [3.27], yet we like to do things in this way as dimensional analysis becomes more straightforward. The presence of measurement units, for example, shows at first glance that we are in the non-normalized case. Note that one should not confuse italic "*s*" in equation [3.24], which is the normalized Laplace variable, with roman "s", which is a measurement unit and stands for seconds in equation [3.27]. Employing a coherent typesetting for equations is not an option here[5].

3.2.7.2. *Fourth-order* 10 kHz *Butterworth low-pass approximation*

The advantage of normalization is that we do not need to redo all the calculations in section 3.2.7.1, but only the denormalization. For example, if we choose $f_c = 10\,\mathrm{kHz}$, the following equations are obtained:

$$s = \frac{p}{2\pi \times 10\,\mathrm{kHz}}, \qquad [3.28]$$

$$H(p) = \frac{1}{\left(\frac{p}{2\pi\times 10\,\mathrm{kHz}}\right)^2 + \frac{1}{1.3058}\frac{p}{2\pi\times 10\,\mathrm{kHz}} + 1} \times \frac{1}{\left(\frac{p}{2\pi\times 10\,\mathrm{kHz}}\right)^2 + \frac{1}{0.54087}\frac{p}{2\pi\times 10\,\mathrm{kHz}} + 1}. \qquad [3.29]$$

3.3. Active filter synthesis by means of standard cells

Once the transfer function $H(p)$ has been written, we can factor it in the form of equation [3.4] and proceed to the synthesis of the different

5 Standards for that exist, such as ISO 80000-2.

terms. We are dealing with low-pass filters, so all circuits presented in this section will have the required low-pass behavior. For an all-pole filter, we therefore need some circuits (cells) able to synthesize:

– a couple of complex-conjugate poles;

– a real negative pole.

Dozens of solutions based on different technologies have been studied for this purpose. Here, we tackle only some basic examples using cells built around operational amplifiers.

3.3.1. *Low-pass Sallen-Key cell: a pair of complex conjugate poles*

For a couple of complex-conjugate poles, a widespread solution is the Sallen–Key filter cell[6], as shown in Figure 3.6. This circuit is part of a class of circuits called EPF from Enhanced Positive Feedback and indeed two feedback loops are present: one via R_3 and R_4 as well as via C_1, the latter being responsible for a certain amount of positive feedback.

Circuit analysis techniques allow to write down the transfer function of the circuit:

$$H(p) = \frac{K}{1 + [R_2C_2 + (1-K)R_1C_1 + R_1C_2]p + R_1R_2C_1C_2p^2}, \tag{3.30}$$

where K is the DC gain of the circuit:

$$K = 1 + \frac{R_3}{R_4}. \tag{3.31}$$

6 Hans Camezind (1934–2012), the designer of the ubiquitous NE555 timer, once wrote: “Analog designers don’t get Nobel prizes, they get a circuit named after them” [CAM 05].

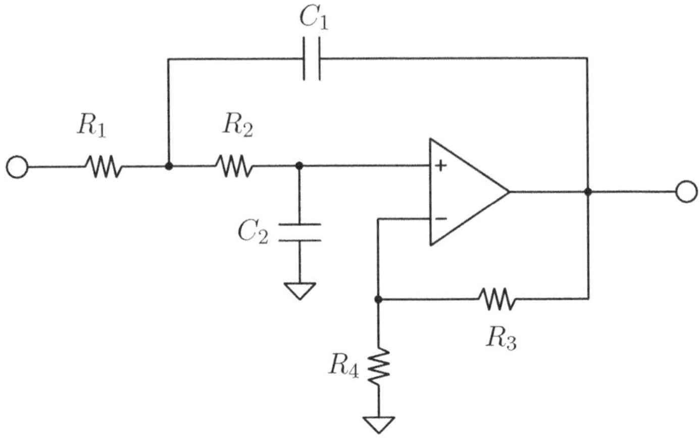

Figure 3.6. *A Sallen–Key second-order low-pass cell. Be sure that the bias current of the non-inverting input of the operational amplifier is provided*

It is, therefore, a sort of double RC filter with an amplifier having a gain K, so certain authors call it a KRC filter [FRA 15]. Sallen–Key cells have a relatively low output impedance (equal to zero for an ideal operational amplifier), so cascaded cells do not change their individual response. Each couple of poles can therefore be synthesized separately, which greatly simplifies the calculations. In comparison, circuits such as the one shown in Figure 3.2 are considerably harder to calculate.

As seen in section 3.2.3, each couple of complex conjugate poles is uniquely identified by two parameters: ω_s and q_s. The circuit in Figure 3.6 contains six passive components. So the problem is indeterminate and additional conditions should be taken into account.

Moreover, the presence of a positive feedback loop indicates that stability constraints exist and the designer should be careful with the choice of the gain K, since $K > 3$ yields instability. $K = 1$ is a popular choice (so $R_4 \to \infty$) in order to optimize the gain-bandwidth product of the circuit. Sometimes, $C_1 = C_2$ or alternatively $R_1 = R_2$ can be taken. Such choices allow the calculation of the remaining components by matching the coefficients of p in [3.30].

Solutions different from Sallen–Key exist (multiple feedback, biquad, Rauch and many others). They offer different trade-offs between sensitivity to device tolerances, bandwidth, number of components, etc.

3.3.2. *Low-pass active RC cell: a real negative pole*

To synthesize a real pole in the transfer function $H(p)$, several different cells exist. Simple ones are depicted in Figure 3.7. Circuit analysis shows that the first one is an inverting cell with a first-order low-pass behavior:

$$H(p) = \frac{-1}{1 + pRC}. \qquad [3.32]$$

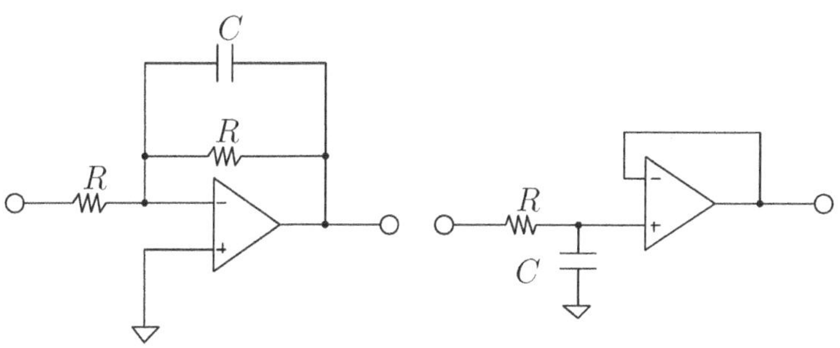

Figure 3.7. *First-order low-pass cells (inverting and non-inverting)*

Values of R and C can be adjusted to obtain the real pole as needed. The two resistors do not need to be identical (the circuit remains a first-order one), introducing a certain amount of gain or attenuation in the band-pass.

The second circuit is instead a simple buffered RC low-pass filter:

$$H(p) = \frac{1}{1 + pRC} \qquad [3.33]$$

Once again, a degree of freedom is left to the designer, as the same value of the pole can be obtained with different choices of R and C.

3.3.3. *Cell order*

Figure 3.8 shows an example of a sixth-order all-pole filter synthesized with three Sallen–Key low-pass cells. By adjusting the values of the passive components, the three responses in Figure 3.4 (Butterworth, Bessel–Thompson, Chebyshev) can be obtained with exactly the same circuit topology.

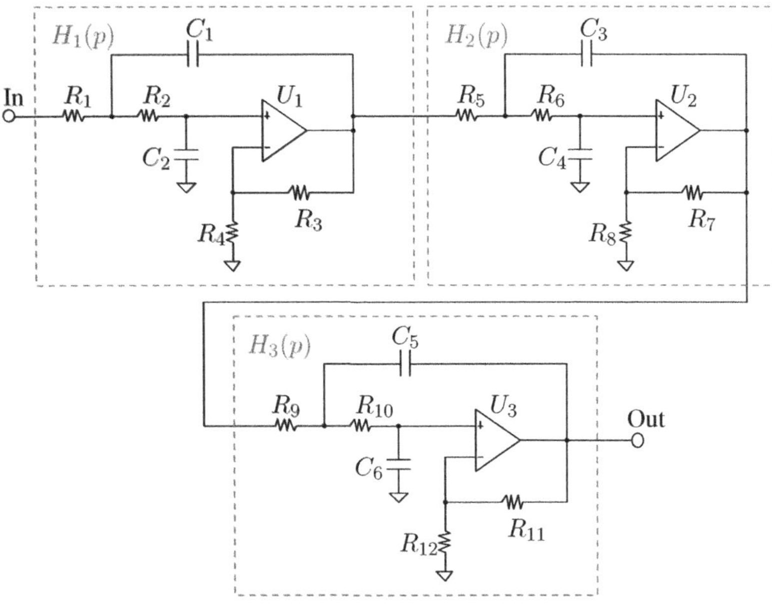

Figure 3.8. *A sixth-order all-pole filter. The overall response is calculated as the product of the transfer functions of the three Sallen–Key cells:* $H(p) = H_1(p) \times H_2(p) \times H_3(p)$, *at least if each cell operates inside its linearity range*

We discuss, for example, a sixth-order Chebyshev-like 1 dB response (one of those represented in Figures 3.4 and 3.5). As we have seen, and as summarized in Table 3.3, we have three complex-conjugate couples of complex poles. By using equation [3.6] for each second-order term, the complete (normalized) Chebyshev transfer function is as follows:

$$H_{\mathrm{n}}(s) = \frac{1}{1.1220} \times \frac{1}{\frac{1}{0.99535^2}s^2 + \frac{1}{0.99535\times 8.0038}s + 1} \times \frac{1}{\frac{1}{0.74681^2}s^2 + \frac{1}{0.74681\times 2.1981}s + 1} \times \frac{1}{\frac{1}{0.35313^2}s^2 + \frac{1}{0.35313\times 0.76086}s + 1}. \qquad [3.34]$$

Their individual response of each factor is shown in Figure 3.9, as well as the overall transfer function [3.34]. We note that, due to a high value q_{s}, the couple of poles named "A" in Table 3.3 and Figure 3.9 exhibits a considerable peaking ($\approx 18\,\mathrm{dB}$) of the response.

Name	Pole s_{s}	ω_{s}	q_{s}	Associate to
A	$-0.06218 + \mathrm{j}0.99341$ $-0.06218 - \mathrm{j}0.99341$	0.99535	8.0038	$H_3(p)$
B	$-0.16988 + \mathrm{j}0.72723$ $-0.16988 - \mathrm{j}0.72723$	0.74681	2.1981	$H_2(p)$
C	$-0.23206 + \mathrm{j}0.26618$ $-0.23206 - \mathrm{j}0.26618$	0.35313	0.76086	$H_1(p)$

Table 3.3. *Poles (normalized frequency) of a $1\,\mathrm{dB}$ Chebyshev approximation*

There is a degree of freedom concerning the order with which the couples of poles should be distributed. If the first cell synthesizing $H_1(p)$ in the circuit in Figure 3.8 is calculated to obtain a "type A" response, peaking might result in a saturation of the output stages of the operational amplifier U_1. This situation is tricky, since the other cells smooth the signal, so that clipping may not be evident by inspecting only the output of U_3. This phenomenon can reduce the overall dynamical range, and therefore a practical rule can be followed: *arrange cells in ascending order of* q_{s}. So, in our example, couple C is attributed to $H_1(p)$, B to $H_2(p)$ and A to $H_3(p)$.

This rule should not be followed when signal amplitude is very small and the main concern is noise and not clipping, but this seldom happens, given the usual position of the filter (i.e. after the amplifier) in the measuring system.

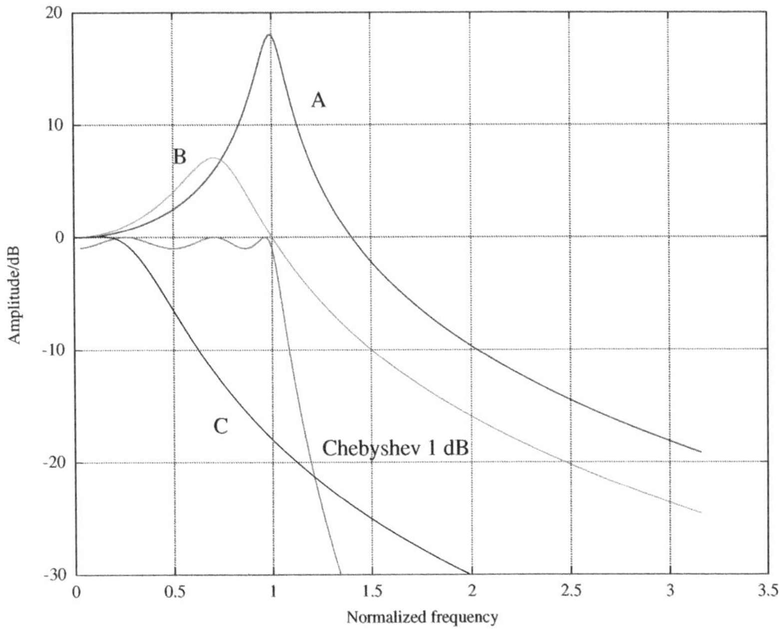

Figure 3.9. *Decomposition of a sixth-order* 1 dB *Chebyshev response into those of individual second-order cells (see Table 3.3)*

3.4. Frequency transform techniques

Up to now, we have exclusively treated low-pass filters. You might wonder if the same theory might be developed for other kinds of filter responses such as high-pass, band-pass and notch. The answer is of course affirmative, but a shortcut exists and consists of reusing all results seen above with low-pass filters, by exploiting frequency transforms. In practice, as we will see in the following sections, every kind of response can be reduced to a standard, equivalent, low-pass behavior by applying an appropriate nonlinear frequency transform.

3.4.1. *High-pass filters*

Figure 3.10 shows a typical request for a high-pass filter. Note how the frequency axis is mirrored if compared with Figure 3.3.

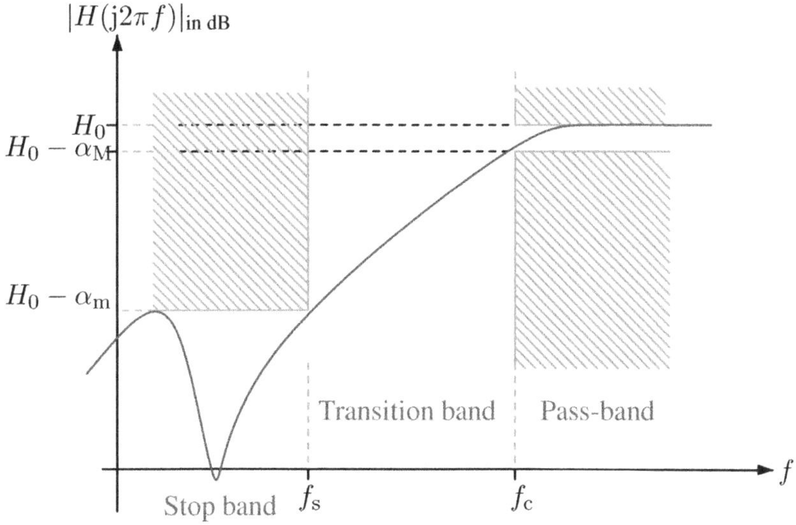

Figure 3.10. *Attenuation behavior requested from a high-pass filter*

Following this observation, the frequency transform to be applied is:

$$s_b = \frac{1}{s}. \qquad [3.35]$$

The complex variables s and s_b are to be interpreted as follows:

– s is the *normalized* complex Laplace variable for the high-pass response;

– s_b is the *normalized* complex Laplace variable for the equivalent low-pass response.

Note that since both variables are normalized, measurement units do not appear in equation [3.35], which is therefore perfectly legal.

3.4.2. *Band-pass filters*

Typical attenuation constraints concerning band-pass filters are shown in Figure 3.11. A particularity of band-pass filters is that there are two rejected bands, and therefore two corners identified by f_{s1}, f_{s2} and α_{m1} and α_{m2}. The band-pass has two limits: f_{c1} and f_{c2}. The frequency transform is therefore a combination of a low-pass and a high-pass filter response. We define two auxiliary quantities:

$$f_0 = \sqrt{f_{c1} f_{c2}} \qquad [3.36]$$

$$Q_0 = \frac{f_0}{f_{c2} - f_{c1}}, \qquad [3.37]$$

where f_0 is a sort of "barycentric" frequency, the geometric average of f_{c1} and f_{c2}, and it is used for the normalization:

$$s = \frac{p}{2\pi f_0}. \qquad [3.38]$$

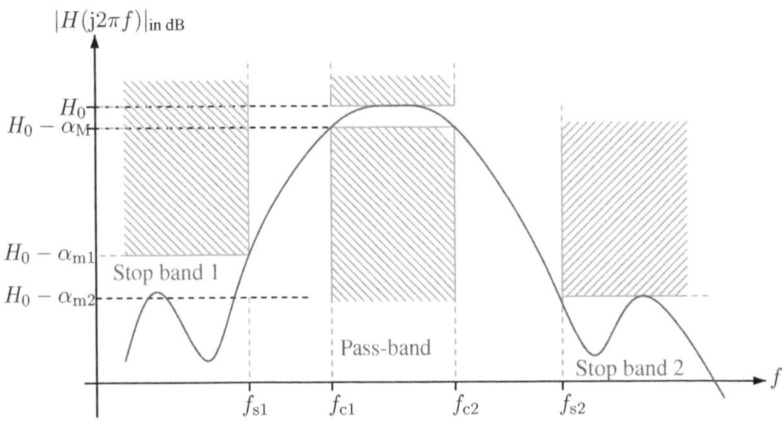

Figure 3.11. *Attenuation behavior requested from a band-pass filter*

On the other hand, the Q_0 factor is a measure of the filter selectivity. High values of Q_0 yield to selective filters.

The complete frequency transform is as follows. Note that a nice symmetric form is obtained when the normalization condition [3.38] is included in the transform:

$$s_b = Q_0\left(s + \frac{1}{s}\right) = Q_0\left(\frac{p}{2\pi f_0} + \frac{2\pi f_0}{p}\right). \qquad [3.39]$$

The constraints coming from the two stop bands are transformed into their low pass equivalents. The filter must be calculated by considering the most restrictive one.

3.4.3. *Band-reject (notch) filters*

A notch filter response is visible in Figure 3.12. Typically, this kind of filters are extremely useful when an unwanted noise is predictable in a certain spectral band that does not contain useful information for the signal.

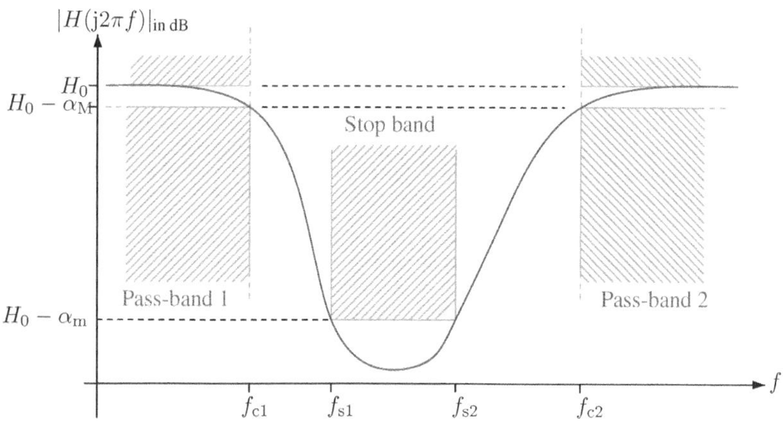

Figure 3.12. *Attenuation behavior for a notch filter*

The frequency transform is symmetric to the one seen in section 3.4.2. We therefore begin to define similar quantities Q_0 and f_0:

$$f_0 = \sqrt{f_{c1} f_{c2}}, \quad [3.40]$$

$$Q_0 = \frac{f_0}{f_{c2} - f_{c1}}. \quad [3.41]$$

The frequency transform is as follows:

$$s_b = \frac{1}{Q_0 \left(s + \frac{1}{s}\right)} = \frac{1}{Q_0 \left(\frac{p}{2\pi f_0} + \frac{2\pi f_0}{p}\right)}. \quad [3.42]$$

3.4.4. *High-pass and band-pass cells*

Frequency transforms yield to approximation of functions having behaviors different from low-pass. In this case, different kinds of cells must be employed for the synthesis of the circuit. Figure 3.13 shows an example of a standard cell that can be employed for the synthesis of a second-order high-pass response.

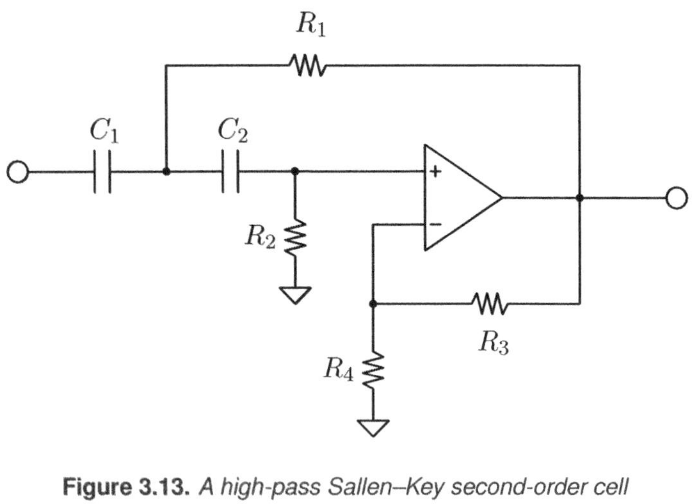

Figure 3.13. *A high-pass Sallen–Key second-order cell*

Low-pass and high-pass responses can be combined in a single second-order cell to achieve a band-pass behavior. This is the case of the Sallen–Key variant shown in Figure 3.14. An example of a simple first-order cell implementing a high-pass response is visible in Figure 3.15.

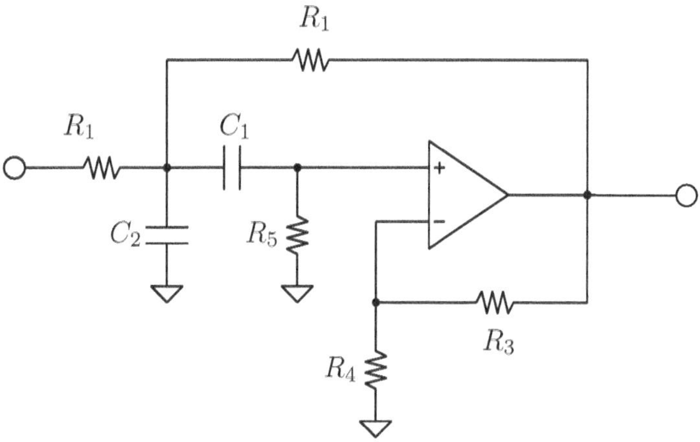

Figure 3.14. *A band-pass second-order cell*

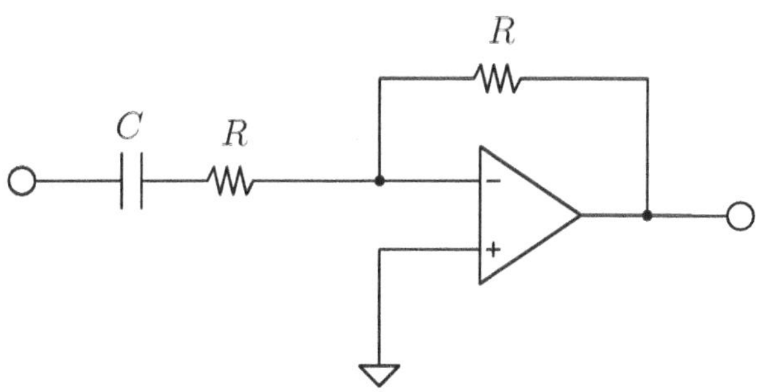

Figure 3.15. *A simple first-order inverting high-pass cell*

As an exercise, the transfer functions of those circuits may be deduced.

3.5. Conclusion

In this chapter, we have introduced the main terminology adopted in filter design and we have briefly described the approximation strategies useful for all-pole filters (Chebyshev, Butterworth and Bessel–Thompson). We saw how the direct synthesis of a passive filter is difficult, but using active cells allows to tackle it via a decomposition in simple low-order terms. What has been seen is by no means a detailed description of the huge domain of filter synthesis. Our goal has mainly been to familiarize with the most frequent problems.

4

Analog to Digital Converters

4.1. Digital to analog converters and analog to digital converters: an introduction

Computers and digital systems are wonderfully powerful when number-crunching tasks are required. Astonishingly, complex operations can be performed in the blink of an eye by cheap microcontrollers. A kid playing with a modern smartphone exploits a computing power and memory far greater than supercomputers employed for Apollo missions to the moon. Digital storage is cheap and compression algorithms have reached a tremendous efficiency. In other words, there is a clear convenience associated with the digital processing of signals.

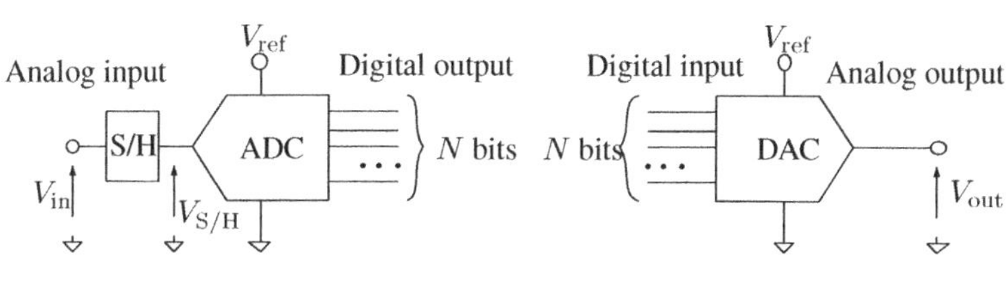

Figure 4.1. *Block diagram representation of analog to digital converter (ADC) and digital to analog converters (DAC)*

Converters represent the link between the analog and digital worlds. An analog to digital converter (ADC) samples an input signal (i.e. takes snapshots of the voltage) and delivers a code on N bits, which digitally represents the sample. A digital to analog converter (DAC) performs the opposite function and converts a code into an analog voltage. Figure 4.1 shows the block diagram symbols, which are usually employed for an ADC and a DAC. In the case of an ADC, it has to perform the following activities:

– the *sampling*, which consists of extracting a sample at a specific moment t_0 yielding an analog voltage that is held. The circuit that performs this function is called a *sample and track* or *sample and hold circuit* (S/H for short), as shown in Figure 4.1. The sampling operation may be repeated at a regular pace at a sampling frequency F_e, thus transforming a continuously varying analog signal into a discrete set of analog voltages. If the Nyquist–Shannon theorem is respected, no information is lost at this stage (see section 3.2.1 for a discussion of aliasing effects)[1];

– the *quantization*, which consists of transforming a sampled voltage into a digital code represented in N bits. The number of codes being a finite number, this operation introduces a quantization error and a possible loss of information;

– the digital code on N bits must be sent to an external digital unit (a microcontroller, a Field Programmable Gate Array (FPGA), a digital bus of some kind...) by means of an appropriate protocol. For the simplest converters with a parallel interface, a certain number of wires are employed for the N bits, as well as control and handshaking lines. More recently, serial interfaces such as Inter-Integrated Circuit (I^2C), Serial Peripheral Interface (SPA) or more complex protocols can be employed to minimize the number of pins in the packaging;

– the control of the converter: to configure it and precisely trigger the instant when a conversion has to be done.

1 Undersampling techniques (i.e. deliberately not following the Nyquist–Shannon condition by choosing a comparatively low F_e) can be employed for the acquisition of a narrowband signal, as long as loss of information due to spectral superposition is accurately avoided. The sampling operation, which cannot be perfectly instantaneous, must still remain fast enough, even if the sampling itself is done at a relatively slow pace.

In the following sections, we will describe some classic strategies to implement the first two functions. The digital communication and control of the converter are however digital interfacing applications and therefore fall outside of the scope of this book.

The main goal of the chapter is to briefly describe the challenges that exist in ADC structures from the point of view of someone who has to choose the right integrated circuit for a particular application. Since ADCs are rather complicated structures, in the past high-performance solutions were expensive hybrid modules. It is indeed an astonishing achievement for the art of crafting modern monolithic analog circuits that converters with impressive characteristics are now available at a very convenient cost. Therefore, we will present the conversion strategies by showing block diagrams, without going into too much detail, which can be found in resources such as [KES 05]. On the other hand, we will spend a certain amount of time discussing the conversion errors introduced by converters and we will present some examples of performances of real devices as they may be found in data sheets.

At the end of the chapter, we will briefly describe the conversion errors that exist and have to be taken into account during the choice of the right conversion strategy.

4.2. Notations and digital circuits

Here we would like to introduce the digital blocks that we will employ in this chapter. Although this is by no means a complete survey, we will employ mixed-signal circuit elements, as shown in Figure 4.2, which deserve to be presented briefly:

– a digital line, which is just a single wire. The voltage with respect to the reference node represents a "logic zero" or a "logic one"[2] if it lies inside intervals defined by the employed logic family (more or less advanced CMOS circuits, the old TTL, etc…);

2 In some cases, it may be useful to differentiate between a "high" logic level and a "one" as well between a "low" and a "zero", but we will not discuss those details here.

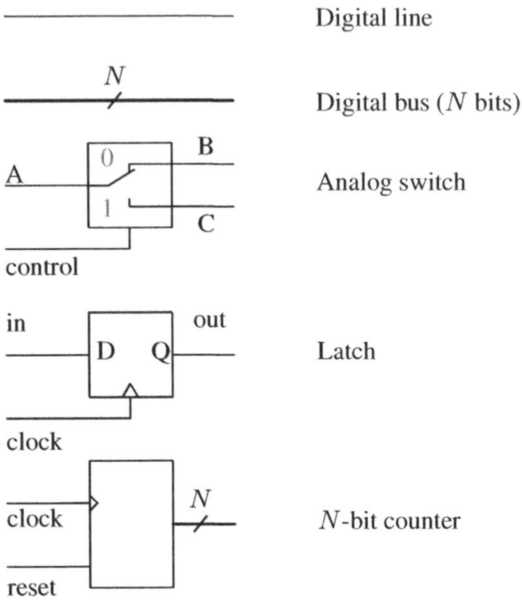

Figure 4.2. *Basic digital or mixed-signal elements used in this chapter*

– a N-bit parallel bus, which is simply made by routing N digital lines along the same path. A connection between a digital line and a bus is shown by means of a little triangle with an indication as to which bit of the bus the connection has to be made with. Of course, with N lines we have N bits and we may represent 2^N different values (from 0 to $2^N - 1$) with the standard binary representation. An alternative representation exists and it is sometimes employed in converters. It is based on the observation that in the binary codes, we need, in certain situations, to flip several bits to move from a value to the successive one. For example, the binary equivalent of $(7)_{10}$ (base 10) is $(0111)_2$ (base 2) where $(8)_{10}$ is $(1000)_2$ so we need to flip the values of all four bits. In some applications, this is not a good thing and instead of adopting the classic binary coding for the digital values, the Gray code is employed. Giving all the details is out of the scope of this introductory book, however the most important property is that in the Gray code, it is sufficient to change one bit to move from a code to the following one.

In the example seen above, 7 (base 10) is represented with 0100 (Gray) where $(8)_{10}$ is 1100 (Gray), only the first bit is changed;

– an analog switch, which connects the analog input A to B when the control line has a digital zero and A to C when the control line has a digital one. Different techniques exist, ranging from diode-based gates or bridges to field effect transistors. Modern CMOS-integrated devices usually advantageously exploit the ohmic region of MOSFETs, which can be switched on and off by modulating the voltage applied to the gate;

– a latch, which is a memory element. Every time there is a clock "tick", the logic level in the D input is transferred to the Q output;

– a counter, which starts from zero at each reset and then increments the number on the output N-bit bus, literally counting the number of cycles of the signal clock.

In general, the clock is quite an important signal in digital electronics, as it represents a synchronization signal that may be used to control circuits. A clock "tick" is a transition, which may be a rising or falling edge in the clock signal.

To represent negative integer values with N bits, several techniques exist but the most commonly employed is the two's complement. Negative numbers are represented with the quantity that should be added to the one to be represented to reach 2^N. For example, on four bits we can represent numbers from -8 to 7 with the two's complement technique. Codes $(0000)_2$ to $(0111)_2$ represent positive numbers with the ordinary binary notation, code $(1111)_2$ represents -1, since if we add $(0001)_2$ to $(1111)_2$ we get $(1\,0000)_2$, which is 2^N, which is of course not representable with only four bits. Code $(1110)_2$ represents -2 and so on.

In the following sections, we will first describe sample and hold circuits followed by different strategies of ADCs. These structures employ the building blocks described here.

4.3. Sample and hold circuits

As we will see in the rest of the chapter, analog to digital conversion requires a certain amount of time during which the voltage to convert is kept as constant as possible. Therefore, a sampling operation is required to "freeze" the signal at a constant value for the appropriate time interval. This is the role of the "sample and hold" (or "sample and track") circuit, which ideally transforms a continuously varying signal V_{in} into a signal $V_{\mathrm{S/H}}$. The latter is equal to $V_{\mathrm{in}}(k/F_{\mathrm{e}})$ in each time interval $k/F_{\mathrm{e}} \leq t < (k+1)/F_{\mathrm{e}}$, where F_{e} is the sampling frequency, as shown in Figure 4.3. Of course, this is what happens in a particular situation when the signal is sampled at a constant frequency F_{e}, but in reality the sample and hold circuit only needs to keep the voltage $V_{\mathrm{S/H}}$ constant for the time needed for the converter to do its job[3]. In practice, the sampling cannot be done instantaneously, but the sample and hold circuit will make a sort of average of the input signal around the sampling time at a time interval τ (being as short as possible). Moreover, a random error in the sampling time (called jitter) can be an important source of conversion noise since it is translated into an amplitude error[4].

Given the need to keep a voltage constant for a given interval, sample and hold circuits are based on an analog memory. This is usually accomplished by means of a capacitor, and the performances of the sample and hold circuit depend on the capacitor itself as well as the capability of charging and discharging it very rapidly when needed. Therefore, a switch and at least two followers are present in the circuit, as shown in the two very simple structures visible in Figure 4.4. In both of them, the storage capacitor is charged by means of the input signal passing through the sampling switch. The output voltage is

3 This is the most frequent application. However, in principle, sampling does not automatically mean that a digital conversion must be done. For example, sampling analog oscilloscopes were used in the past for HF applications. The Tektronix 1S1 sampling head (1 GHz) from 1965 was an exquisite example of such an application based on the undersampling of a narrowband signal.

4 Jitter and τ are usually the factors limiting the applicability of undersampling.

present at the output of an operational amplifier, which avoids the capacitor discharging, apart with the inevitable bias currents.

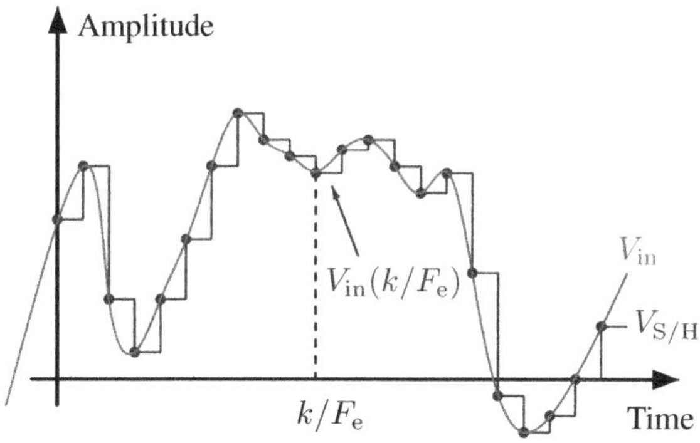

Figure 4.3. *Ideal function of a sample and hold circuit: it samples the input signal V_{in} and keeps it constant until a new sample is required. Circles represent signals at the sampling moments*

The structure at the top is the simplest to understand and the storage capacitor C is charged by means of the input buffer during the sampling time, when the switch S is closed. Apart from simplicity, this strategy offers limited advantages with respect to the topology shown at the bottom, an integrating closed-loop sample and hold. In this circuit, when the switch S is closed, the capacitor C is charged so that the output matches the input. It is possible to employ a high-precision operational amplifier for the input buffer and an operational amplifier with very low bias currents for the integrator.

While the principle of the circuit seems simple, given the circuit constraints and many practical difficulties, designing high-performance sample and hold circuits is a very difficult task. However, in the context of modern ADCs, sample and hold circuits are now almost invariably integrated along the ADCs and the difficulties have been tackled in the design stage of the chip. Imperfections of sample and hold circuits may include gain errors, linearity issues, offsets and jitter in the sampling time. The sampling cannot be instantaneous and one must consider the acquisition time (limited by the amplifiers' slew rate) and the delay time required for the output voltage to stabilize within a given tolerance.

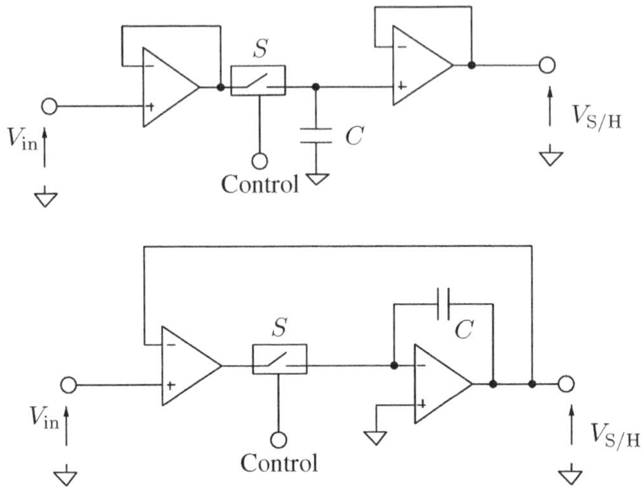

Figure 4.4. *Simple sample/hold structures: (top) open loop, (bottom) closed loop, integrating*

4.4. Converter structures

4.4.1. *General features*

With respect to the operations described in section 4.1, several parameters qualify the behavior of the converters and deserve to be discussed. For example, one of the most important points to consider is the resolution, and how the values are represented at the output of the ADC (or at the input of a DAC). If we employ N bits, a discrete set of 2^N different codes is available. The number of bits therefore gives important information on the resolution of the converter. In fact, for example, a hypothetical 4-bit converter employs 16 different codes to represent the analog signal and Figure 4.5 shows a possible way that this is done.

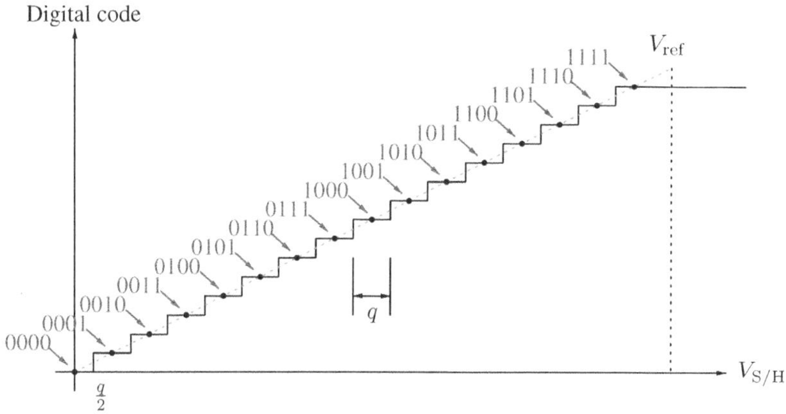

Figure 4.5. *Correspondence between analog values and its digital representation in a 4-bit analog to digital converter with thresholds centered, so that the quantization error is minimized. The configuration shown represents an unipolar converter with full scale at* V_{ref}

In the same context, several other things can be noted. First of all, the full-scale value of the analog voltage is called V_{ref}, which is an analog reference voltage that is required by the converter. Often, V_{ref} is obtained by means of an appropriate circuit: when high accuracy is mandatory, constraints become very stringent on drift, ageing and stability of V_{ref}. In some cases, especially for less delicate applications, V_{ref} is obtained by means of voltage reference integrated with the converter itself or even from the power supply. A second important choice is the position of the thresholds (i.e. the position of the vertical lines in the "steps" in Figure 4.5). In this case, the first threshold is placed at a voltage $q/2$, where q (quantum) is the width of the intervals associated with a change in the last significative bit:

$$q = \frac{V_{\text{ref}}}{2^N}. \qquad [4.1]$$

In this way, the quantization error due to the finite number of digital values is always (in module) less or equal to $q/2$ when $V_{S/H}$ ranges from 0 to $V_{\text{ref}} - q/2$.

The graph in Figure 4.5 represents the ideal behavior of a so-called “unipolar” converter, where analog voltages are always positive and binary codes are unsigned. Other possibilities exist: for example, there are the “bipolar” converters allowing the analog voltage to be positive or negative and therefore employing digital codes with sign. Moreover, the relation between the reference voltage and the full scale may be different from the one shown in Figure 4.5. For example, unipolar converters exist where the full-scale voltage is $2V_{\mathrm{ref}}$ while bipolar converters can have a positive full-scale voltage of V_{ref} and a negative full-scale voltage of $-V_{\mathrm{ref}}$.

To select the best converter for our application, the first page of the data sheet usually summarizes some important characteristics. Relevant information about the number of bits employed, maximum conversion speed, the unipolar/bipolar operation, digital coding and interface is usually given there. Yet, be very very careful: data there may be affected by a certain degree of advertisement hype! Typical values give some orders of magnitude, yet good designs yield circuits able to cope with the most restrictive operating conditions, and therefore are based on minimum or maximum *guaranteed* characteristics (and not the typical ones).

Therefore, one must dig deeper in the data sheets to determine the conversion principle and the exact conditions in which the characteristics have been measured. For example, when the number of bits is high (more than 16), one should be aware that the device noise floor may limit the number of bits, which is extremely important. Usually, in this case the number of “noise free bits” is specified in the data sheet, it may be non-integer and usually strongly depends on the conversion conditions.

4.4.2. *Flash ADCs*

A flash converter is probably the simplest ADC structure to understand. Let us start our description from a single comparator, which can be seen as a 1-bit ADC, as shown in Figure 4.6.

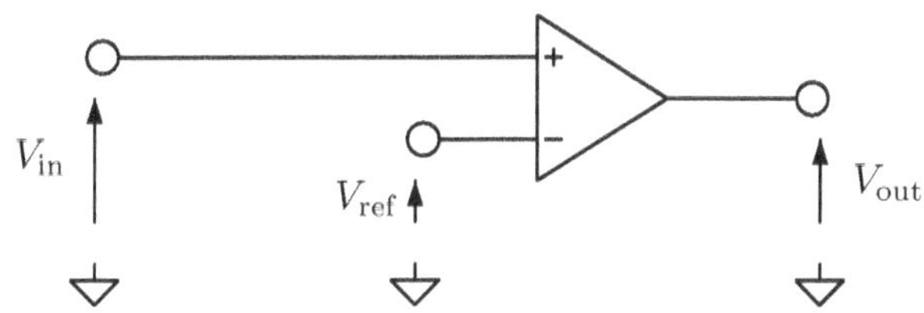

Figure 4.6. *A simple comparator. This is not an operational amplifier even if the symbol is the same*

Don't be fooled by the fact that the symbol of a comparator is the same as the one of an operational amplifier (see, for example, Figure 2.1). An operational amplifier is expected to work in its linear region with a negative feedback circuit; a comparator is made to work without it or with a positive one. It is a nonlinear component, which provides an output voltage corresponding to a "1" logic level if $V_{S/H} > V_{ref}$ and one corresponding to a "0" otherwise. If we want to increase the number of bits, the straightforward strategy may be to adopt a certain number of comparators and a resistance ladder to provide the appropriate thresholds. The outputs of the comparators form a digital code often called the "thermometer code". Figure 4.7 shows the circuit of a 3-bit flash converter and Table 4.1 shows the coding matrix, which must be implemented to have an output represented as a binary number $b_2 b_1 b_0$. It is easy to detect an over scale error or a faulty situation.

A flash converter can be very fast (hence its name), since the conversion speed is limited by the reaction time of a single comparator, as all of them operate in parallel. However, the number of bits available is clearly limited by the maximum number of comparators and resistances that can be employed, as they grow exponentially with the number of bits N. Such structures are usually integrated on a single silicon chip and passive devices tend to occupy a lot of surface, so in practice, quite often, most of the chip area is occupied by the resistance ladder. On the other hand, it is relatively easy to achieve a good match between integrated resistances. The logic circuitry needed to implement the decoding function is usually relatively inexpensive and does not constitute a limiting factor for the performances. A second problem is instead that the $V_{S/H}$ voltage is applied at the same time to all comparator inputs, which tied together may represent a hard to drive load (usually mainly capacitive), thus requiring a robust amount of buffering.

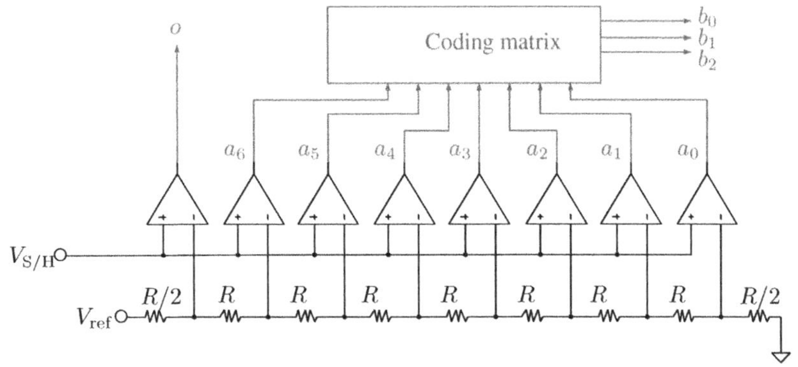

Figure 4.7. *A 3-bit flash converter, thus requiring* $2^3 = 8$ *comparators and* $2^3 + 1 = 9$ *resistances*

a_0	a_1	a_2	a_3	a_4	a_5	a_6	o	b_2	b_1	b_0	Note
0	0	0	0	0	0	0	0	0	0	0	
1	0	0	0	0	0	0	0	0	0	1	
1	1	0	0	0	0	0	0	0	1	0	
1	1	1	0	0	0	0	0	0	1	1	
1	1	1	1	0	0	0	0	1	0	0	
1	1	1	1	1	0	0	0	1	0	1	
1	1	1	1	1	1	0	0	1	1	0	
1	1	1	1	1	1	1	0	1	1	1	
1	1	1	1	1	1	1	1	1	1	1	Over scale error!
All others combinations								x	x	x	Error or fault!

Table 4.1. *Coding matrix for the three-bit flash converter shown in Figure 4.7*

To summarize, flash converters are generally very fast but limited to a low number of bits and therefore relatively inaccurate. A solution based on the divide and conquer strategy is possible and it is only marginally slower: it is called the "half flash converter", but we need first to discuss at least one structure of a DAC since we will need it.

4.4.3. *A simple DAC: R2R ladder*

We give here a description of a simple DAC. Of course, different strategies exist but we have chose to describe the classic solution shown in Figure 4.8 based on a ladder formed by resistances whose value is R or $2R$ (hence the name R-2R or R2R). First of all, a certain number of analog switches are configured following the bits of the binary code, which has to be converted.

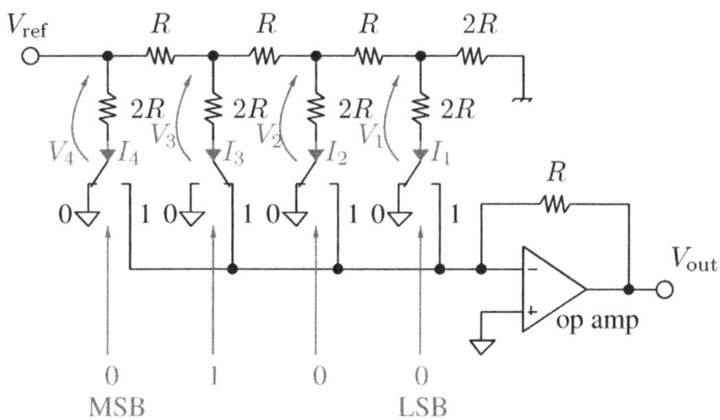

Figure 4.8. *An R2R ladder analog to digital converter. Switches are configured so that the converted binary number is* $(0100)_2$*. An operational amplifier is employed as a current to voltage converter*

To understand how the circuit works, we note that the negative feedback around the operational amplifier forces the voltage to zero on its inverting input, which is thus a virtual ground. Consequently, the currents I_1 - I_4 are unaffected by the position of the switches, as the voltage on the reference node is the same as the one on the inverting input of the operational amplifier. The switches, thus, enable the selection of which currents given by the R2R ladder have to be diverted and summed up on the zero-impedance node. The operational amplifier, wired as a current to voltage converter (or trans-resistance:

see Figure 1.15), transforms the sum of currents into a voltage V_{out} (negative if $V_{\text{ref}} > 0$).

Let us calculate the values of the currents I_1 - I_4 with respect to the value of the resistances R and the reference voltage V_{ref}. We start from I_1 by supposing that V_1 is known:

$$I_1 = \frac{V_1}{2R}. \tag{4.2}$$

Then, we calculate V_1 from V_2 by means of a voltage divider rule. As mentioned, the position of the switch does not affect anything on the ladder, therefore:

$$V_1 = V_2 \frac{2R//2R}{R + 2R//2R} = V_2 \frac{R}{R + R} = \frac{1}{2} V_2. \tag{4.3}$$

When we try to calculate V_2, we note that the resistances combine in the same way by obtaining:

$$V_2 = V_3 \frac{R}{R + R} = \frac{1}{2} V_3. \tag{4.4}$$

This relation can be easily generalized and we note that the currents in the switches follow a power-of-2 law. Therefore, if the switches are controlled by the bits of the binary code to convert, we get a voltage V_{out} that is proportional to its value.

The conversion speed is limited by the commutation time of the switches, as well as the bandwidth and the settling time of the operational amplifier.

4.4.4. *Half-flash and pipeline ADCs*

If we come back to the flash ADC discussed in section 4.4.2, we may be able to imagine splitting the problem in half. Let us suppose that we need a 8-bit converter. In this case, a flash solution would require 256 comparators and a voltage divider composed of 257 resistances. The idea is to employ a first flash converter for the identification of the most significative four bits (16 comparators and 17 resistances), then refine the conversion a second time by adding a second flash converter for the least significative four bits. The block diagram of a half-flash converter is shown in Figure 4.9. The input voltage $V_{S/H}$ is first roughly converted by the upper ADC, giving the four most significative bits of the code. Then, a DAC is employed to obtain the analog voltage corresponding to that first rough conversion. The difference between the output of the DAC and the input voltage is the quantization error of the first converter, which is employed to obtain the least four significative bits by a second ADC.

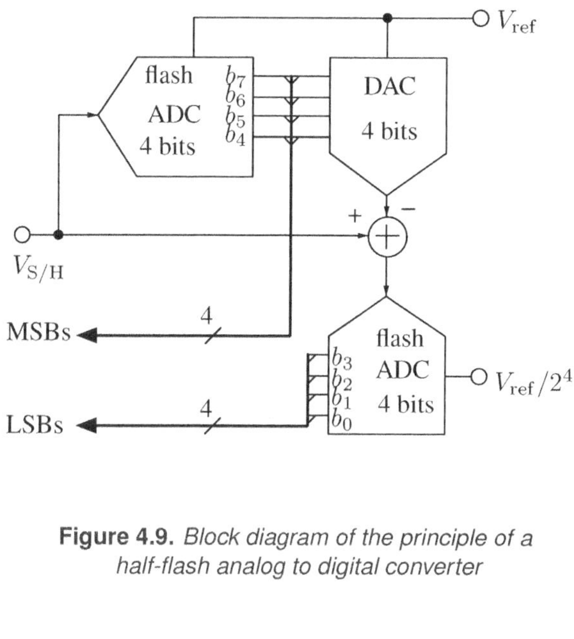

Figure 4.9. *Block diagram of the principle of a half-flash analog to digital converter*

The half-flash ADCs may represent a good trade-off between the complexity of the circuit (much simpler than that of a flash converter) and the maximum sampling speed, which is of course marginally reduced. We can split the number of bits into three or four groups and employ the same idea, obtaining the so-called pipelined converters. The pipeline concept may also be convenient to reduce the conversion time as we do not need every converter to work at the same time, as long as we can afford a certain latency in the output data. Pipelined structures reign over the realm of high-speed and high-accuracy converters.

4.4.5. *Successive approximation converters*

Let us again consider the half-flash converter described in section 4.4.4. The idea was to split the number N of bits in smaller sets. Pipeline converters operate in a "parallel" way by employing several converters active at the same time, each one converting a bunch of bits. We may develop a more drastic version of this idea if we take N single-bit ADCs. We have already employed a simple comparator such as the one shown in Figure 4.6 as a 1-bit ADC[5]. To simplify the circuit, instead of wiring together a pipeline of N comparators with N DACs, we adopt a single comparator and a single DAC by means of a "serial" operation. The leading role here is played by the special N-bit register called the successive approximation register (SAR for short) where the approximation of the converted value is constructed step by step. A block diagram of the converter is shown in Figure 4.10. We may describe the conversion algorithm by means of the following C-style pseudo-code:

```
SAR=0;                  // Set all bits of SAR to zero
m=N-1;                  // Start from MSB of SAR
while (m>=0) {          // Continue for all bits of SAR
    SAR[m]=1;           // Set the m-th bit to one.
    wait(ADC_READY);
    if (V_DAC>V_in)     // Compare SAR results to Vin
        SAR(m)=0;       // Too much!
    m--;                // next bit.
}
// End of the conversion. SAR contains the result to latch.
```

With the notation `SAR[m]`, we access the *m*th bit of the SAR, with the convention that `SAR[`$N-1$`]` is the most significative bit, whereas

5 Quoting Jim Williams: *Strictly speaking, this viewpoint is correct. It is also wastefully constrictive in its outlook. Comparators don't "just compare" in the same way that op amps don't "just amplify"* [WIL 85].

SAR[0] is the least significative bit. We indicated with `wait(ADC_READY);` a pause needed for the DAC to complete the conversion, usually an appropriate number of cycles of the clock signal controlling the circuit.

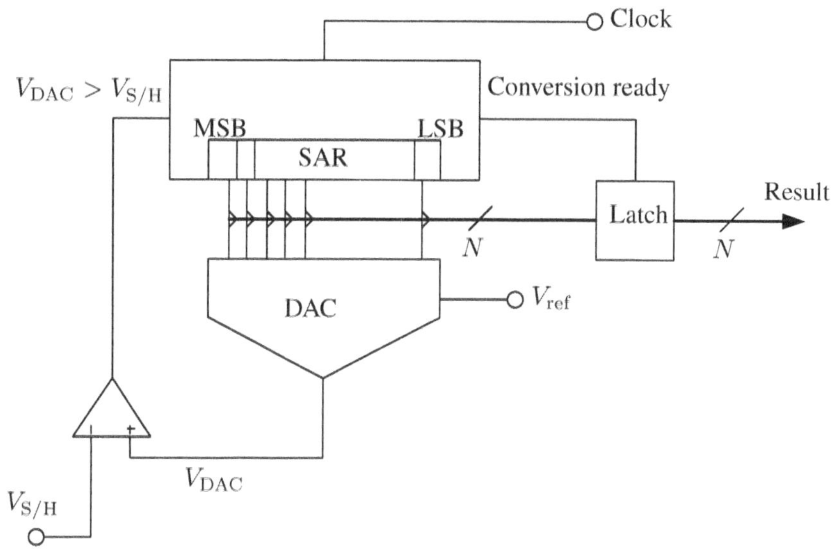

Figure 4.10. *Block diagram of the principle of a successive approximation analog to digital converter*

Most books and descriptions employ the term "successive approximation register" to indicate both the SAR itself, along with the control logic described in the pseudo-code given above. We do not need a particularly complex logic to implement the flow. A simple state machine will be more than enough.

Since the result of the conversion is constructed step by step and is changing, when the conversion is ready, a latch allows it to "freeze" the result for the outside world.

4.4.6. *Single- and double-ramp converters*

If there is something that is easy to do in digital electronics, it is counting to measure time intervals. Therefore, if we can transform the input voltage into such an interval, we may measure it to obtain the conversion result. This is the idea behind single- and double-ramp converters. It is easy to see that if the time interval we are measuring is not very short, one can easily obtain a very high measurement resolution. Therefore, this conversion strategy is very efficient for those cases where the counter employs a high number of bits and a low conversion speed is required. Long occupying first place in the pool of solutions available in this context, they are now progressively being replaced by sigma-delta converters, as described below.

A very simple single-ramp acquisition system is shown in Figure 4.11. The reference voltage V_{ref} is fed to an integrator composed of an operational amplifier, the resistance R and the capacitor C, which charges linearly with a slope equal to:

$$\frac{\mathrm{d}V_{\mathrm{r}}}{\mathrm{d}t} = -\frac{V_{\mathrm{ref}}}{RC}. \qquad [4.5]$$

Therefore, the time between the moment when switch S is opened and the moment when V_{r} becomes equal to $V_{\mathrm{S/H}}$ is proportional to $V_{\mathrm{S/H}}$ itself:

$$\tau = \frac{RC}{-V_{\mathrm{ref}}} V_{\mathrm{S/H}} \qquad [4.6]$$

and measuring that time digitally by means of a counter yields the conversion result.

This method, called single-ramp analog to digital conversion (using a Wilkinson converter, from Sir Denys Haigh Wilkinson), works well but is affected by a potential error coming from drifts in the values of R and C. However, this type of converter can easily be extended to multiple channels by sharing the same ramp generator and counter, as shown in Figure 4.12. If we do not want to see the output buses continuously changing during the conversion because of the counters operation, of course, they must be latched appropriately.

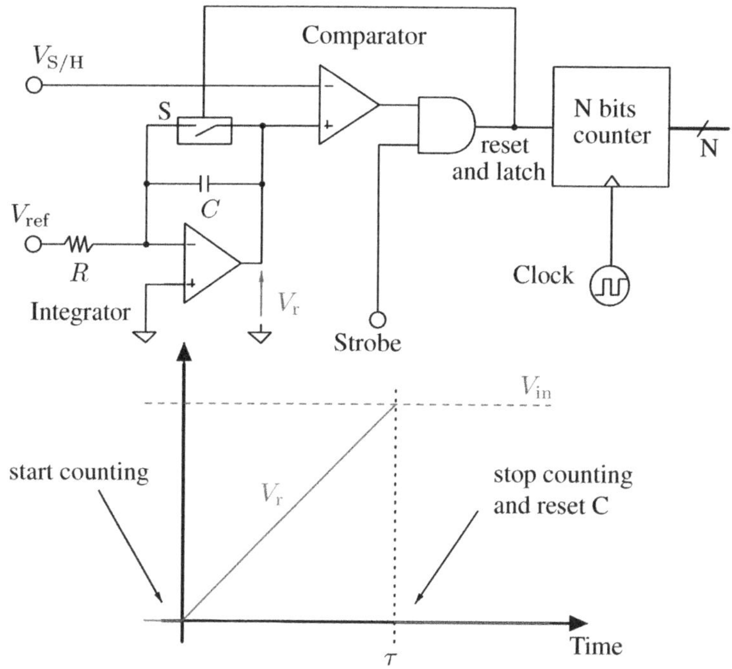

Figure 4.11. *Block diagram of the principle of a single-ramp (Wilkinson) analog to digital converter. A positive slope such as the one shown is obtained with this circuit by means of a negative* V_{ref}

An alternative strategy ensuring that the result does not depend on passive components such as R and C is called double-ramp conversion. A possible implementation is shown in Figure 4.13. Here, the conversion is done in two steps:

– at first, S_1 is closed and S_2 is open. The capacitor C is, therefore, charged with a current $V_{S/H}/R$ and the slope of V_r is:

$$\frac{dV_r}{dt} = -\frac{V_{S/H}}{RC}. \quad [4.7]$$

The duration of this phase (controlled by the glue logic) is constant and corresponds to the time needed for the N-bit counter to count from

0 to $2^N - 1$. Then, the voltage reached is thus:

$$V_{p1} = 2^N T \frac{dV_r}{dt} = -2^N T \frac{V_{S/H}}{RC}. \quad [4.8]$$

where T is the period corresponding to one clock cycle:

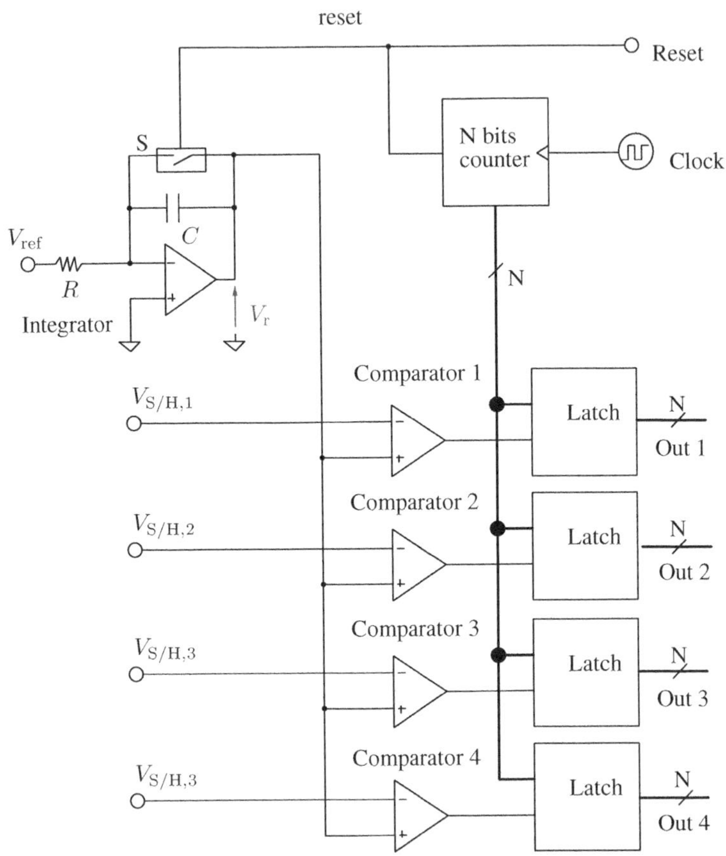

Figure 4.12. *Block diagram of the principle of a multiple channel single-ramp converter*

– in the second phase, S_1 is open and S_2 is closed. The capacitor C is, therefore, discharged with a current V_{ref}/R. The slope is thus:

$$\frac{\mathrm{d}V_r}{\mathrm{d}t} = -\frac{V_{ref}}{RC}. \quad [4.9]$$

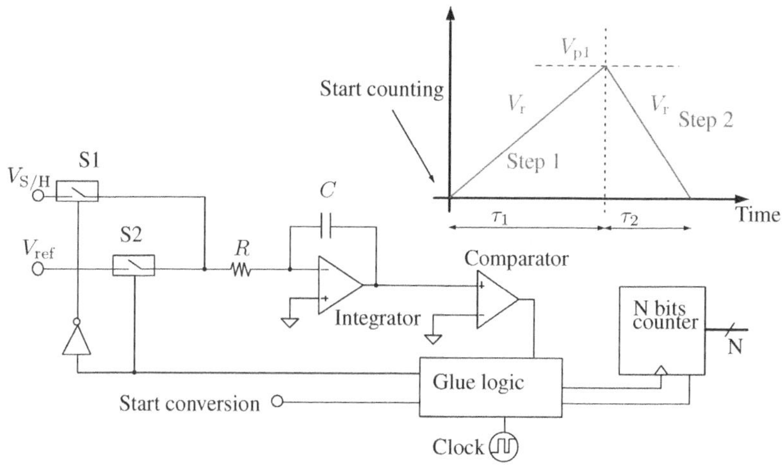

Figure 4.13. *Block diagram and timing diagram of a double-ramp converter*

The duration of this phase is not constant, and it is thus proportional to the voltage reached at the end of the first phase:

$$\tau_2 = V_{p1}\left(\frac{\mathrm{d}V_r}{\mathrm{d}t}\right)^{-1} = -V_{p1}\frac{RC}{V_{ref}} = 2^N T\frac{V_{S/H}}{V_{ref}}. \quad [4.10]$$

Therefore, if the counter counts during this phase, as long as $|V_{S/H}| < |V_{ref}|$ it will not overflow and will reach a value proportional to: $V_{S/H}/V_{ref}$, yielding the result of the conversion.

The main advantage of the double-ramp converter is that there is not a direct dependence on values of passive devices such as R and C. Long-term variations on these components will be cancelled out because of the self-calibrating nature of the converter.

4.4.7. *Sigma-delta converters*

As we saw previously, it is easy and convenient to employ comparators as crude yet fast 1-bit ADCs, so let us also explore the opposite operation. A 1-bit DAC is just a controlled switch that puts 0 V at its output to convert a logic zero and V_{ref} to convert a logic one. The sigma-delta (or delta-sigma) converter principle consists of employing such simple elements in a system with feedback, where the difference between the input value and the result of a 1-bit conversion is integrated, as shown in Figure 4.14. Successively, the bit flow is digitally filtered and decimated (i.e. the sampling frequency is reduced).

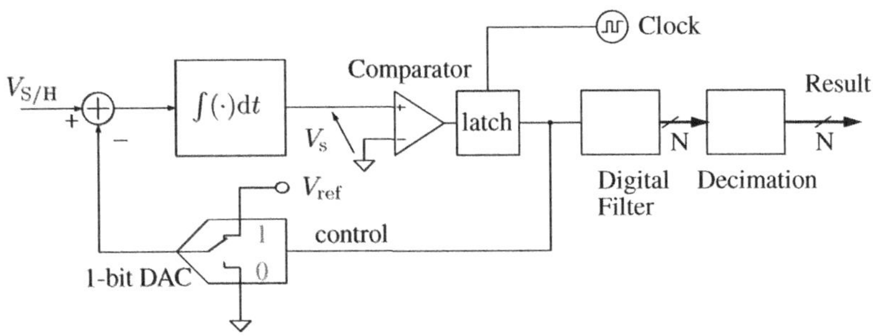

Figure 4.14. *The basic block diagram of a sigma-delta analog to digital converter. The block with the integral sign is an integrator*

To understand how the strategy works, let us put things in the following way. At first, the circuit shown in Figure 4.14 converts $V_{S/H}$ caught between 0 V and V_{ref}. Let us suppose that the output of the DAC is zero), a positive voltage $V_{S/H}$ is integrated, until the output of the integrator becomes positive and thus the comparator outputs a one. At the next clock tick, the 1-bit DAC therefore connects its output to V_{ref}. At the input of the integrator, we thus will have a negative voltage $V_{S/H} - V_{ref}$ which, once integrated, will tend to decrease the voltage at the input of the comparator. If $V_{S/H}$ is large, the output of the integrator becomes rapidly positive and large and we have to integrate several clock ticks in the negative $V_{S/H} - V_{ref}$ voltage to change the polarity of V_s. We thus have a one followed by several zeros at the output. Vice versa, if $V_{S/H}$ is close to 0 V, we obtain sequences of a single one followed by a lot of zeros.

We obtain a regular flow of ones and zeros and we may measure their frequency. The number of ones in the total number of bits will be proportional to $V_{S/H}$, whereas the number of zeros will be proportional to $V_{ref} - V_{S/H}$. Since the sum of the two is equal to the total number of bits, we may calculate the proportion of ones in an explicit way:

$$P_1 = \frac{V_{S/H}}{V_{S/H} + V_{ref} - V_{S/H}} = \frac{V_{S/H}}{V_{ref}} \quad [4.11]$$

To summarize, we obtain a very crude 1-bit representation of the signal indicating that the signal has increased or decreased, but we can also count the number of bits in a certain time interval (or implement a moving average mechanism), integrating these changes and increment the resolution of the conversion. This latter kind of operation is a filtering operation that can be implemented digitally. The result of the filtering is a stream of converted codes represented with N bits but at a greatly reduced pace with respect to the clock frequency. The operation of reducing the sampling frequency is called decimation. There is no risk of aliasing in this operation, since it is done after the digital filter has been applied. The most important property of the sigma-delta converters is without doubt its noise shaping properties. In fact, oversampling is made extremely convenient by the structure of the converter, which spectrally shifts the conversion noise out of the bandwidth of the useful signal.

The example described above is the so-called first-order sigma-delta converter and certain blocks such as the summing node and integrator are practically implemented by means of a switched capacitor technique. More complicated higher order structures exist which allow us to obtain even better performances, in particular for what concerns noise shaping characteristics.

Sigma-delta converters can be implemented by means of very simple and efficient circuits. Digital filters are easy to implement and decimation is also convenient. However, the implicit 1-bit operation is based on oversampling, and therefore the final sampling frequency cannot be extremely high. On the other hand, the oversampling means that the requirements for the analog anti-aliasing filter (see Chapter 3) are greatly relaxed, and that is a definite advantage of this very interesting converter topology. Sigma-delta converters are now so convenient that they are employed for the vast majority of applications where high accuracy and low speed is required. More about this can be found in [KES 09], which also includes a historical perspective on such converters.

4.5. No silver bullet: choosing the best trade-off

4.5.1. *Conversion errors and artifacts*

ADCs introduce errors in the measurement chain. Some are inevitable, such as the quantization errors. Some, on the other hand, depend on the device and structure chosen and are specified in the data sheet. We describe the most important ones in Figure 4.15:

– Figure 4.15(a): Offsets introduced during the conversion: static errors that shift the curve shown in Figure 4.5 horizontally. The error is therefore the same for all digital codes in the plot.

– Figure 4.15(b): The differential linearity error, which is associated with a change in the length of certain "steps" in the conversion staircase. In some cases, a large linearity error may even render the conversion curve not monotonic or produce missing codes.

– Figure 4.15(c): The integral linearity error, which is the result of an accumulation of differential linearity errors, concurring together in distorting the line connecting the centers of the steps of the staircase.

– Figure 4.15(d): Gain errors for which the overall conversion gain of the converter is not the one expected.

Linearity errors are static errors, yet they affect dynamic measurements. For example, if very pure sine waves are acquired, nonlinearities introduce harmonics and intermodulation products in the digital signal. By means of a discrete Fourier transform, parameters such as the total harmonic distortion and the spurious free dynamic range can be measured. The measurement challenges can be impressive in this regard, especially with high resolution converters (see, for example, [WIL 11]).

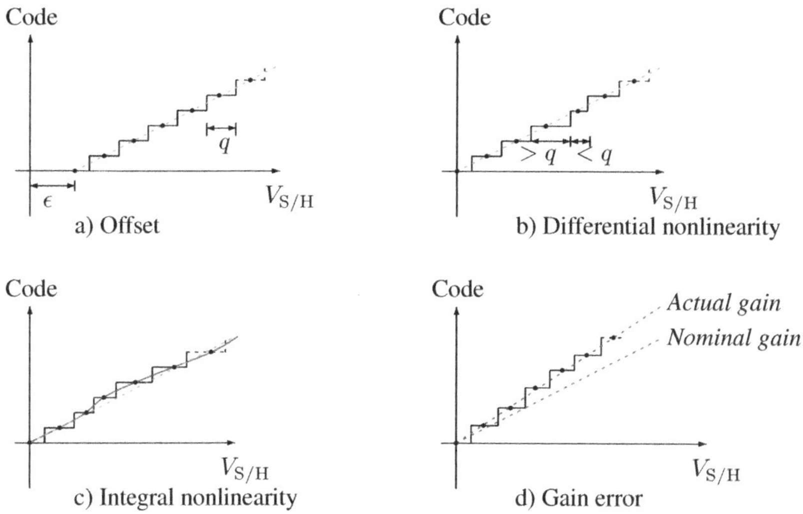

Figure 4.15. *Graphical representation of analog to digital converter errors*

The offsets and gain errors can be corrected by means of a calibration. However, temperature drifts remain and their effects may affect the result in high-resolution converters. It is also important to take into account the errors and the drifts present in the reference voltage provided to the converter as we saw that *every* converter needs a reference. Precise integrated reference sources (bandgap or buried zener) exist and are often available along with the converter itself but have a certain degree of error and thermal drifts. For this reason, some high-performance converters include a temperature measurement system: temperature drifts can thus be corrected digitally.

4.5.2. *Performances of typical converters*

Figure 4.16 shows the first page of a family of simple low-cost ADCs, ADC121C02x. We learn straight away that they are based on a successive approximation strategy (thus something similar to that described in section 4.4.5) on 10 or 12 bits. The communication on the digital side of the converter (i.e. the control commands as well as the result of the conversion) is done by means of a bidirectional synchronous serial protocol, which is called I^2C. This protocol requires only one line for the clock (SCL) and one line for the data (SDA). Once the converter is instructed to launch a conversion (specific codes exist for that), the 10 or 12 bits are sent serially one by one on the SDA line. Therefore, the package can be very compact and convenient and the so-called TSOT (standing for Thin Small Outline Transistor package) is indeed very small, yet still something we can solder manually.

We know that approximation converters are not particularly fast; however, the maximum sampling frequency is 188.9 kSPS (kilo samples per seconds), which is more than enough for audio applications. Each sample (10 or 12 bits) is packed in 2 bytes, sent through the I^2C port, therefore this performance is probably limited more by the maximum speed of the serial bus in the high-speed mode (3.4 MHz) than by the performance of the converter itself. In fact, this is confirmed by the 1 μs time needed for the conversion which suggests that the converter itself could run marginally faster with a higher-speed digital interface. Note how the data sheet proudly states that there are "no missing codes". These converters employ the power supply rail as the reference voltage.

Figure 4.17 shows a radically different beast. The LTC2209 is a *very* fast 16-bit converter employing a pipeline converting principle. The declared sampling frequency is 160 MSPS and the overall performances are quite impressive. In this case, the converted data are made available in a parallel bus to sustain the massive data throughput at full speed. The analog input is differential, which among other things helps to reduce noise and harmonic distortion.

National Semiconductor

January 17, 2008

ADC121C021/ADC121C027 I²C-Compatible, 12-Bit Analog-to-Digital Converter (ADC) with Alert

ADC121C021/ADC121C027
I²C-Compatible, 12-Bit Analog-to-Digital Converter (ADC) with Alert Function

General Description

The ADC121C021 is a low-power, monolithic, 12-bit, analog-to-digital converter(ADC) that operates from a +2.7 to 5.5V supply. The converter is based on a successive approximation register architecture with an internal track-and-hold circuit that can handle input frequencies up to 11MHz. The ADC121C021 operates from a single supply which also serves as the reference. The device features an I²C-compatible serial interface that operates in all three speed modes, including high speed mode (3.4MHz).

The ADC's Alert feature provides an interrupt that is activated when the analog input violates a programmable upper or lower limit value. The device features an automatic conversion mode, which frees up the controller and I²C interface. In this mode, the ADC continuously monitors the analog input for an "out-of-range" condition and provides an interrupt if the measured voltage goes out-of-range.

The ADC121C021 comes in a small TSOT-6 package with an alert output. The ADC121C027 comes in a small TSOT-6 package with an address selection input. The ADC121C027 provides three pin-selectable addresses. Pin-compatible alternatives are available with additional address options.

Normal power consumption using a +3V or +5V supply is 0.26mW or 0.78mW, respectively. The automatic power-down feature reduces the power consumption to less than 1μW while not converting. Operation over the industrial temperature range of −40°C to +105°C is guaranteed. Their low power consumption and small packages make this family of ADCs an excellent choice for use in battery operated equipment.

The ADC121C021 and ADC121C027 are part of a family of pin-compatible ADCs that also provide 8 and 10 bit resolution. For 8-bit ADCs see the ADC081C021 and ADC081C027. For 10-bit ADCs see the ADC101C021 and ADC101C027.

Features

- I²C-Compatible 2-wire Interface which supports standard (100kHz), fast (400kHz), and high speed (3.4MHz) modes
- Extended power supply range (+2.7V to +5.5V)
- Up to four pin-selectable chip addresses
- Out-of-range Alert Function
- Automatic Power-down mode while not converting
- Very small 6-pin TSOT packages
- ±8kV HBM ESD protection (SDA, SCL)

Key Specifications

- Resolution — 12 bits; no missing codes
- Conversion Time — 1μs (typ)
- INL & DNL — ±1 LSB (max) (up to 22kSPS)
- Throughput Rate — 188.9kSPS (max)
- Power Consumption (at 22kSPS)
  - 3V Supply — 0.26 mW (typ)
  - 5V Supply — 0.78 mW (typ)

Applications

- System Monitoring
- Peak Detection
- Portable Instruments
- Medical Instruments
- Test Equipment

Pin-Compatible Alternatives

All devices are fully pin and function compatible.

Resolution	ALERT Output	ADDR Input
12-bit	**ADC121C021**	**ADC121C027**
10-bit	ADC101C021	ADC101C027
8-bit	ADC081C021	ADC081C027

Connection Diagrams

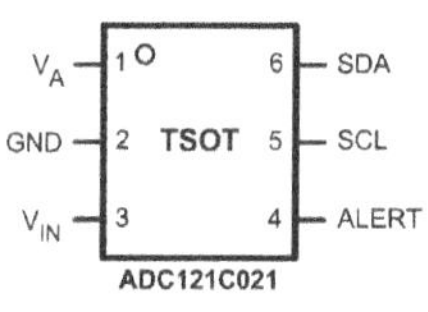

30020901

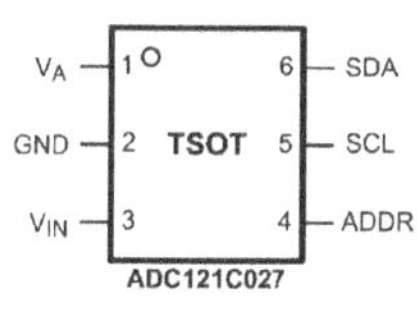

30020902

I2C™ is a registered trademark of Phillips Corporation.

 www.national.com

Figure 4.16. *The first page of the data sheet of ADC121C0xx converters. Courtesy of Texas Instruments*

LINEAR TECHNOLOGY

LTC2209

16-Bit, 160Msps ADC

FEATURES

- **Sample Rate: 160Msps**
- **77.3dBFS Noise Floor**
- **100dB SFDR**
- **SFDR >84dB at 250MHz (1.5V_{P-P} Input Range)**
- **PGA Front End (2.25V_{P-P} or 1.5V_{P-P} Input Range)**
- **700MHz Full Power Bandwidth S/H**
- **Optional Internal Dither**
- **Optional Data Output Randomizer**
- LVDS or CMOS Outputs
- Single 3.3V Supply
- Power Dissipation: 1.53W
- Clock Duty Cycle Stabilizer
- Pin-Compatible Family:
  130Msps: LTC2208 (16-Bit), LTC2208-14 (14-Bit)
  105Msps: LTC2217 (16-Bit)
- 64-Pin (9mm × 9mm) QFN Package

APPLICATIONS

- Telecommunications
- Receivers
- Cellular Base Stations
- Spectrum Analysis
- Imaging Systems
- ATE

DESCRIPTION

The LTC®2209 is a 160Msps 16-bit A/D converter designed for digitizing high frequency, wide dynamic range signals with input frequencies up to 700MHz. The input range of the ADC can be optimized with the PGA front end.

The LTC2209 is perfect for demanding communications applications, with AC performance that includes 77.3dBFS Noise Floor and 100dB spurious free dynamic range (SFDR). Ultra low jitter of $70fs_{RMS}$ allows undersampling of high input frequencies with excellent noise performance. Maximum DC specs include ±5.5LSB INL, ±1LSB DNL (no missing codes).

The digital output can be either differential LVDS or single-ended CMOS. There are two format options for the CMOS outputs: a single bus running at the full data rate or demultiplexed busses running at half data rate. A separate output power supply allows the CMOS output swing to range from 0.5V to 3.6V.

The ENC$^+$ and ENC$^-$ inputs may be driven differentially or single-ended with a sine wave, PECL, LVDS, TTL or CMOS inputs. An optional clock duty cycle stabilizer allows high performance at full speed with a wide range of clock duty cycles.

TYPICAL APPLICATION

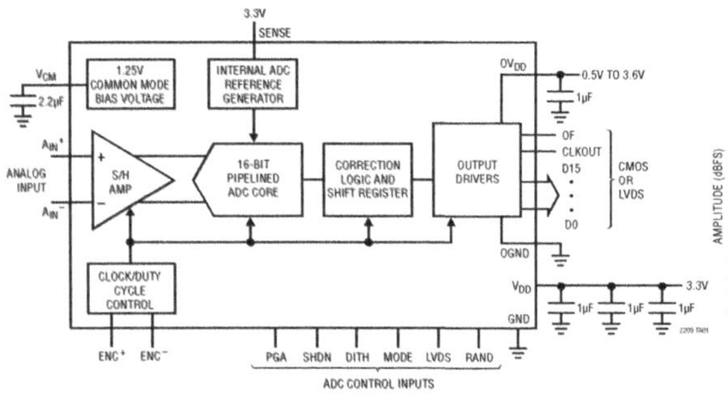

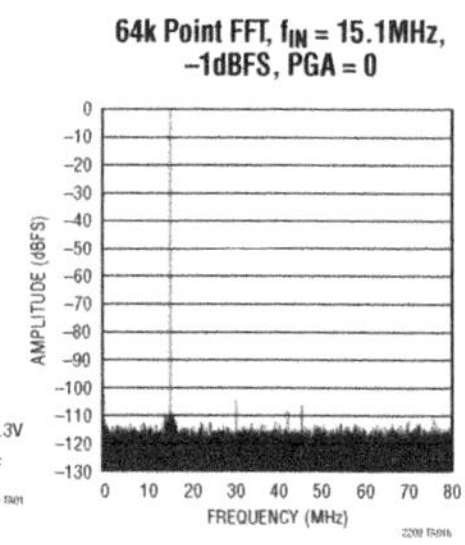

2209fb

Figure 4.17. *An extract from the data sheet of LTC2209 converter. Courtesy of Linear Technology*

ANALOG DEVICES

32-Bit, 10 kSPS, Sigma-Delta ADC with 100 µs Settling and True Rail-to-Rail Buffers

Data Sheet **AD7177-2**

FEATURES

32-bit data output
Fast and flexible output rate: 5 SPS to 10 kSPS
Channel scan data rate of 10 kSPS/channel (100 µs settling)
Performance specifications
- **19.1 noise free bits at 10 kSPS**
- **20.2 noise free bits at 2.5 kSPS**
- **24.6 noise free bits at 5 SPS**
- **INL: ±1 ppm of FSR**

85 dB filter rejection of 50 Hz and 60 Hz with 50 ms settling
User configurable input channels
- **2 fully differential channels or 4 single-ended channels**
- **Crosspoint multiplexer**

On-chip 2.5 V reference (±2 ppm/°C drift)
True rail-to-rail analog and reference input buffers
Internal or external clock
Power supply: AVDD1 − AVSS = 5 V, AVDD2 = IOVDD = 2.5 V to 5 V
- **Split supply with AVDD1/AVSS at ±2.5 V**

ADC current: 8.4 mA
Temperature range: −40°C to +105°C
3- or 4-wire serial digital interface (Schmitt trigger on SCLK)
- **Serial port interface (SPI), QSPI, MICROWIRE, and DSP compatible**

APPLICATIONS

Process control: PLC/DCS modules
Temperature and pressure measurement
Medical and scientific multichannel instrumentation
Chromatography

GENERAL DESCRIPTION

The AD7177-2 is a 32-bit low noise, fast settling, multiplexed, 2-/4-channel (fully/pseudo differential) Σ-Δ analog-to-digital converter (ADC) for low bandwidth inputs. It has a maximum channel scan rate of 10 kSPS (100 µs) for fully settled data. The output data rates range from 5 SPS to 10 kSPS.

The AD7177-2 integrates key analog and digital signal conditioning blocks to allow users to configure an individual setup for each analog input channel in use. Each feature can be user selected on a per channel basis. Integrated true rail-to-rail buffers on the analog inputs and external reference inputs provide easy to drive high impedance inputs. The precision 2.5 V low drift (2 ppm/°C) band gap internal reference (with output reference buffer) adds embedded functionality to reduce external component count.

The digital filter allows simultaneous 50 Hz and 60 Hz rejection at a 27.27 SPS output data rate. The user can switch between different filter options according to the demands of each channel in the application. The ADC automatically switches through each selected channel. Further digital processing functions include offset and gain calibration registers, configurable on a per channel basis.

The device operates with a 5 V AVDD1 supply, or with ±2.5 V AVDD1/AVSS, and 2 V to 5 V AVDD2 and IOVDD supplies. The specified operating temperature range is −40°C to +105°C. The AD7177-2 is available in a 24-lead TSSOP package.

FUNCTIONAL BLOCK DIAGRAM

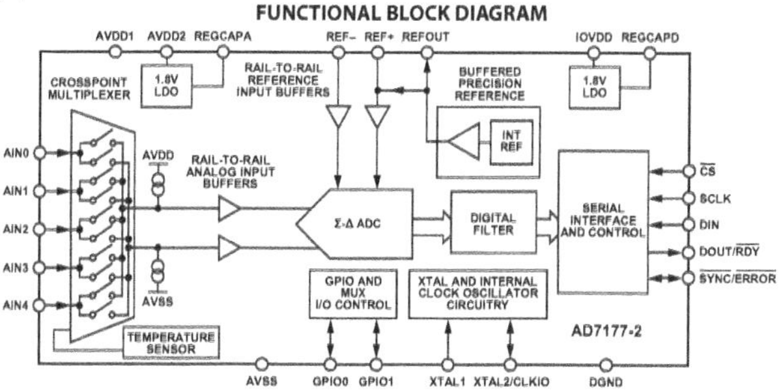

Figure 1.

Rev. B Document Feedback

One Technology Way, P.O. Box 9106, Norwood, MA 02062-9106, U.S.A.
Tel: 781.329.4700
Technical Support www.analog.com

Figure 4.18. *The first page of the AD7177-2 converter data sheet (source: Analog Devices)*

A look at the first page of the AD7177-2 sigma-delta converter shown in Figure 4.18 is not for the feint of heart. However, as always in these cases, there are trade-offs and physical constraints to consider. As we saw in section 4.4.1, even if it is a "32-bit" converter, the data sheet states that the number of "noise free bits" is less than 32 and depends on the sampling speed, which has been adopted. This is rather typical of high-resolution converters and means that the output code fluctuates slightly. In this region of ultrahigh resolution, 1 ppm represents roughly 12 LSBs if the samples are represented on 24 bits. Temperature drifts, gain and linearity errors may easily introduce errors decreasing the number of significative bits in a practical, real-world situation. It is not strange, in this context, that the device includes a temperature sensor: thermal effect can be compensated digitally in a calibration phase. In general, achieving more than 20 bits of absolute accuracy would be the reward of extremely careful temperature and drift compensation. Extreme care should be dedicated to providing clean, well decoupled, power supply voltages to the different parts of the chip (AVDD1, AVDD2 for the analog parts and IOVDD for the digital part) and keeping the analog and digital reference pins separate (AVSS and DGND).

4.6. Conclusion

In this chapter, we briefly described some of the classical problems associated with analog to digital conversion. We tried to give an overall description useful for those who have to choose a converter for a specific application. At first, we briefly described the challenges of the sampling and hold circuits, the first element in the signal path. Then, we detailed some converting structures. As usual in engineering[6], trade-offs must be applied. A converter may be very fast (flash, half flash or pipelined) or very accurate (double ramp or sigma-delta), but both characteristics are very hard (and expensive) to achieve at the same time. Trade-offs exists such as the successive approximation converter. When power consumption becomes an issue, such as for

6 And in real-life problems...

battery-operated applications, it is also important to choose a frugal converter. We have not discussed DACs very much apart from the R2R converter, since it is often used in the structures of ADC we presented. We briefly discussed conversion errors and we presented several examples of typical converters.

5

Introduction to Noise Analysis in Low Frequency Circuits

Be not afeard; the isle is full of noises,
Sounds and sweet airs, that give delight and hurt not.
Sometimes a thousand twangling instruments
Will hum about mine ears, and sometime voices
That, if I then had waked after long sleep,
Will make me sleep again: and then, in dreaming,
The clouds methought would open and show riches
Ready to drop upon me that, when I waked,
I cried to dream again.

William SHAKESPEARE, The Tempest

5.1. What is noise?

It is not easy to give a unique answer to that simple question. In fact, depending on the context, noise can be studied from a number of quite different points of view. However, the presence of noise constitutes one of the most important limitations for the performances of an analog circuit. No analysis on a measurement chain is therefore complete if it does not consider how noise affects performances. We will therefore adopt here a very pragmatic point of view, which will be enough for the purposes of an introductory book.

In the previous chapters, we followed the signal path from the sensor to the analog to digital converter. In our description, the signal was carrying useful information. However, along the processing chain, the signal might be distorted and some unwanted perturbations are inevitably added. It turns out that, at least in our cruel world, it is impossible to fabricate a circuit in which the signal is processed but not perturbed in any measurable way. A certain degree of information is therefore lost in the process. *We call noise everything that is present along the signal without carrying useful or exploitable information.* Therefore, noise tends to mask the useful signal and entails a loss of information. For this reason, the signal-to-noise ratio (SNR) plays an important role and it is often expressed in dB:

$$\mathrm{SNR}|_{\mathrm{dB}} = 10\log\left(\frac{P_{\mathrm{signal}}}{P_{\mathrm{noise}}}\right), \qquad [5.1]$$

where P_{signal} is the average power of the signal (i.e. what carries useful information) and P_{noise} is the average power of the noise present. An electronic circuit treats signals in a certain *dynamic range*, whose lower limit is determined by the minimum acceptable value of the SNR and whose upper limit by linearity issues (such as clipping), for high-amplitude signals. Most noise calculations require to understand how the result of equation [5.1] is affected by a particular circuit. We can roughly distinguish between two different kinds of noise:

– noise coming from the measurement chain itself, even operated alone. This is the case of stochastic noise generated inside the chain.

– noise coming from circuits placed externally to the measurement chain itself, perturbing it in some way. This is an electromagnetic compatibility (EMC) issue, dealing with unwanted coupling.

The distinction is often quite questionable, for example when there are internal couplings inside the measurement chain itself. For instance, the large amplitude signal at the output of the amplifier might perturb the input. However, the distinction helps us, at least in an initial approach to the problem. *However, there is no such thing as an infinite SNR*. Signal and noise are always present together, given the physical and technological constraints and the environment of the circuit. We may more appropriately ask for each situation what is the tolerable amount of SNR.

Noise is often represented with stochastic models; this chapter begins with concepts and definitions useful for handling stochastic finite-power signals. Of course, far from being a complete course in signal processing (see, for example, [GAR 90] for the detailed mathematical treatment of random processes), only some basic concepts are given. We then describe several kinds of noises, which are present in analog circuits, as well as low-frequency noise models for operational amplifiers. At the end of the chapter, we will provide a very short introduction of the issues related to EMC.

5.2. Stochastic modeling of a noise

5.2.1. *Some definitions*

5.2.1.1. *Stochastic processes as a family of statistical samples*

A stochastic noise is a signal that can be represented by means of a stochastic process, which is a *probabilistic model of a set of waveforms*. Let us imagine, for instance, to acquire three times the hiss noise $B(t)$ coming from a FM radio out of tune by means of a recording oscilloscope. We would obtain three different results, as shown in Figure 5.1. The three different experiences in the picture are indicated by the different tags s_1, s_2, s_3. Of course, we do not expect to have the same result every time, yet we would like to define a certain number of parameters that allow us to say something about the statistical properties of such a signal. For example, if we consider a time instant t_0 measured with respect to the instant we begin recording each sequence, we obtain three different *samples* $B(t_0, s_1)$, $B(t_0, s_2)$, $B(t_0, s_3)$ for the three times we repeat the experience. We may raise questions such as *what is the probability that, if we repeat the experience a fourth time, the sample* $B(t_0, s_4)$ *will lie inside a given interval?* Or *what can we say about frequency contents of the signal?* We give some definitions and try to give enough mathematical instruments and tools to answer these questions.

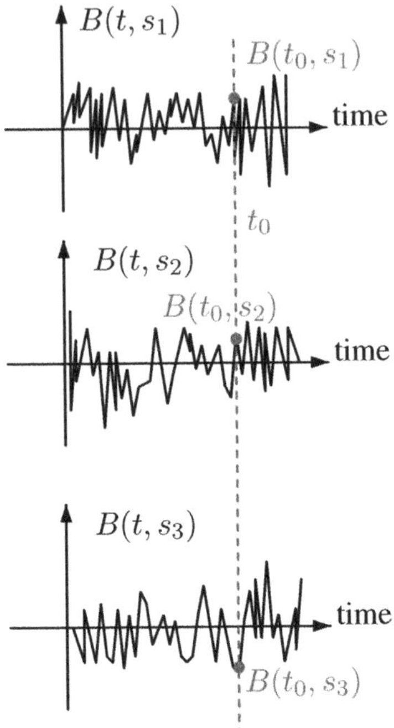

Figure 5.1. *Three different statistical samples of a single stochastic continuous-time process: noise* $B(t)$

5.2.1.2. *Probability density distribution and power spectral density*

Given what was said in the previous section, we may first observe in Figure 5.1 "vertically" by choosing a particular time t_0. The process $B(t_0, s)$ thus is a stochastic variable. We can employ a probability density distribution $p_B(b)$ yielding, at a given time t_0, to the integral:

$$\mathbb{P}_{b_1,b_2} = \int_{b_1}^{b_2} p_B(b)\mathrm{d}b \qquad [5.2]$$

which gives the probability that, if we do the experience and we take a sample at a time t_0, it is caught between b_1 and b_2. Moreover, the probability density distribution is usually normalized, such that:

$$\mathbb{P}_{-\infty,+\infty} = \int_{-\infty}^{+\infty} p_B(b)\mathrm{d}b = 1 \qquad [5.3]$$

In our description, we will always suppose that our processes are *stationary in the wide sense*, stating that their average calculated with equation [5.4] does not depend on the instant t_0:

$$\mu_{t_0} = \int_{-\infty}^{+\infty} b p_B(b)\mathrm{d}b = \mathbb{E}\{B(t_0)\} = \mu, \qquad [5.4]$$

where the notation $\mathbb{E}\{B(t_0)\}$ is a shortcut for the probability integral, the *expected value* of $B(t_0)$. That definition of stationarity also requires that the autocovariance function depends only on the difference between the two instants t_1 and t_2 chosen for the calculation:

$$C_B(t_1, t_2) = \mathbb{E}\{(B(t_1) - \mu_{t_1})(B(t_2) - \mu_{t_2})\} = C_B(\tau, 0), \qquad [5.5]$$

where $\tau = t_1 - t_2$. The wide-sense stationarity is a reasonable hypothesis if the stochastic process is employed to modelize a physical system whose properties does not change in time. We are, moreover, interested in signals whose average μ is equal to zero.

The second important observation is related to a "horizontal" analysis of what happens, as shown in Figure 5.1, by trying to quantify the "rate of change" of signal variations. To be more precise, we need to quantify precisely in which portions of the spectrum power is carried. This tool[1] is the power spectral density $\Gamma_B(f)$. Given a certain frequency range $f_1 - f_2$, the power carried by the signal in the range is proportional to the variance, calculated by the integral:

$$\sigma^2_{f_1,f_2} = \int_{f_1}^{f_2} \Gamma_B(f)\mathrm{d}f \qquad [5.6]$$

1 We will not give the details here. As we will see later, it will be enough to know that $\Gamma_B(f)$ is the Fourier transfer of the autocorrelation function $\gamma_B(\tau)$, which is calculated in the time domain (this requires the wide-sense stationarity).

If we calculate the variance by integrating on the whole spectrum, we obtain the square of the root mean square (RMS) amplitude of the signal:

$$b_{\mathrm{RMS}}^2 = \sigma^2 = \int_{-\infty}^{+\infty} \Gamma_B(f)\mathrm{d}f \qquad [5.7]$$

5.2.1.3. *Ergodicity*

Another important definition is the *ergodicity*, a property of certain signals that allows to determine their statistical parameters from a time analysis, or, equivalently, from a statistical point of view. For example, the average μ can be calculated from the probability density function as done in equation [5.4]. We can also calculate a time average (the upper bar is a shortcut for the time integral) on, let us say, $B(t, s_1)$:

$$\mu = \overline{B(t, s_1)} = \lim_{T \to +\infty} \frac{1}{T} \int_{-T/2}^{T/2} B(t, s_1)\mathrm{d}t. \qquad [5.8]$$

In an ergodic signal, the two calculations give exactly the same results. In other words, on an ergodic signal, we can calculate the same properties by means of a specific observation in the time domain or by choosing an instant t_0 in time and calculating statistical parameters: reading Figure 5.1 horizontally or vertically yields exactly the same results.

For what concerns the RMS amplitude of $B(t)$, it can be calculated from, let us say, $B(t, s_2)$:

$$b_{\mathrm{RMS}} = \sqrt{\overline{B^2(t, s_2)}} = \sqrt{\lim_{T \to +\infty} \frac{1}{T} \int_{-T/2}^{T/2} B^2(t, s_2)\mathrm{d}t}, \qquad [5.9]$$

or from the probability density:

$$b_{\mathrm{RMS}} = \sqrt{\mathbb{E}\{B^2\}} = \sqrt{\int_{-\infty}^{+\infty} B^2 p_B(b)\mathrm{d}b}. \qquad [5.10]$$

With an ergodic process, the two approaches yield the same results, which is also the same obtained with equation [5.7]. We will consider that all stochastic noise processes discussed in the following sections are ergodic. If we want a classic example of a signal that is *not* ergodic, think about a Gaussian noise that is sampled at an instant t_s and held constant. A simple observation for $t > t_s$ would lead to a constant value, which will give no information about the statistical properties of the sample noise. Yet, the stochastic nature of the signal is still present since each time we repeat the experience, the sampled value will be different.

5.2.2. Measurement units for $p_B(b)$ and $\Gamma_B(f)$

In the context of signal processing, most signals are normalized to simplify dealing with measurement units. The definitions we gave in section 5.2.1 remain coherent with those seen in signal processing courses and books, but we will always deal with signals that are *not* normalized and we must deal correctly with measurement units:

– if $B(t)$ is a voltage signal[2], the measurement unit of a sample taken at a time t_0, i.e. $B(t_0)$ will be a voltage, therefore measured in volt. From equation [5.2], we determine that the measurement unit of $p_B(b)$ is V^{-1};

– if $B(t)$ is a current signal, the measurement unit of a sample taken at a time t_0, i.e. $B(t_0)$ will be a current, therefore measured in ampere. From equation [5.2], measurement unit of $p_B(b)$ is A^{-1}.

For what concerns $\Gamma_B(f)$, we deduce from equation [5.6] that $[\Gamma_B(f)] = [B]^2/\mathrm{Hz}$ where the notation [...] indicates the operator "measurement unit of". However, in practice:

– if $B(t)$ is a voltage signal, the instantaneous power $P(t)$ dissipated on a load resistance R is $P(t) = B^2(t)/R$. Therefore:

$$B(t) \propto \sqrt{RP(t)} \tag{5.11}$$

– if $B(t)$ is a current signal, the instantaneous power dissipated on a load resistance R is $P(t) = RB^2(t)$, so:

$$B(t) \propto \sqrt{P(p)/R} \tag{5.12}$$

2 A signal where the information is directly carried by (i.e. is analogous to) the voltage.

Therefore, when the load resistance R is unknown, for voltage signals the square root of the spectral density is specified, measured in $\mathrm{V}/\sqrt{\mathrm{Hz}}$. Conversely, for current signals, from equation [5.12] we obtain the square root of the spectral density, measured in $\mathrm{A}/\sqrt{\mathrm{Hz}}$. Those are *not* real spectral densities, but their square *has a real physical meaning*. In equation [5.6], the proportionality constant between $\sigma^2_{f_1,f_2}$ and the power in the same band is $1/R$ or R if the signal is, respectively, a voltage or a current signal.

5.2.3. *Negative and positive frequencies*

It is well known from signal processing that the Fourier representation of a signal contains positive as well as negative frequencies. In our case, the power spectral density $\Gamma_B(f)$ is the Fourier transform of $\gamma_B(\tau)$, the autocorrelation function of, for example, $B(t, s_1)$:

$$\gamma_B(\tau) = \lim_{T\to+\infty} \frac{1}{T} \int_{-T/2}^{T/2} B(t, s_1) B^*(t-\tau, s_1)\mathrm{d}t \qquad [5.13]$$

$$\Gamma_B(f) = \int_{-\infty}^{+\infty} \gamma_B(\tau) \exp(-\mathrm{j}2\pi f\tau)\mathrm{d}\tau \qquad [5.14]$$

In practice, the result of a Fourier transform is a complex signal $\Gamma_B(f)$ that features negative as well as positive frequencies. However, when $B(t, s_1)$ is a real signal (that can be observed with an oscilloscope), it can be shown that its autocorrelation function $\gamma_B(\tau)$ is real and symmetric $\gamma_B(\tau) = \gamma_B(-\tau)$. The Fourier transform of a real and symmetric signal has the good taste of being real and symmetric at its turn. *Therefore, knowing and manipulating only positive frequencies is enough to know and manipulate all the information we need*[3].

3 We sometimes read that negative frequencies are only a mathematical trick. I personally do not see that in this way: they are of course intrinsically associated to the Fourier transform, but they represent a generalization of the concept of frequency. They might not play a determinant role here, but there *are* cases where they are indispensable.

Figure 5.2 provides an example of this trick: it is a representation of the calculation of an integral on the complete power spectral density $\Gamma(f)$, which is done on the frequency range (f_1, f_2), as well as on the symmetric interval $(-f_2, -f_1)$. This frequency range is what is obtained, for example, by applying a real bandpass filter on the original signal. Because of the symmetries, we can forget about the negative frequencies, employ only the positive ones and make calculations on $2u(f)\Gamma(f)$ (where $u(f)$ is the Heaviside step function) without having to mess with the symmetric interval $(-f_2, -f_1)$. This approach is so useful and widespread (this is what in practice we did in every example of Chapter 3) that the $2u(f)$ term is often silently left out from the power spectral density.

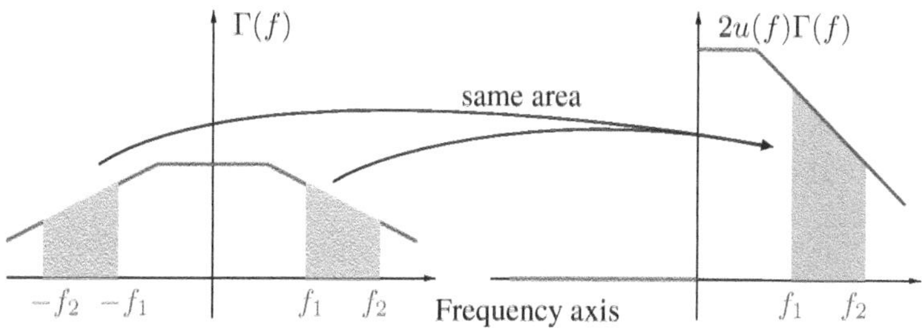

Figure 5.2. *Power calculation on a symmetric power spectral density, originating from a real signal. The two approaches give the same area, but the one at the right deals only with positive frequencies*

5.3. Different kinds of stochastic noises

We call "stochastic noise" a kind of noise that can be mathematically modeled employing a stochastic process. In the following sections, we will review different kinds of noise, giving some information about their physical origin.

If a wheel spins in one direction, we might count the number of revolutions per second in one direction and represent it with a number which is a positive frequency. It makes a lot of sense to use a negative frequency to represent the revolutions per minute in the *opposite* direction.

5.3.1. *Thermal noise (Johnson–Nyquist)*

Thermal noise is due to the thermal agitation of carriers (electron or holes) in a conductor or semiconductor. It is a white noise contribution (its power spectral density is constant) and it is present every time there is a power dissipation in an electronic device. For this reason, as an important term of noise in resistors, and it cannot be avoided, being intrinsically related to the Joule effect. Thermal noise is also called Johnson noise, or Nyquist noise, from the name of John Bertrand Johnson (1887–1970) and Harry Nyquist (1886–1976) who studied it at the Bell labs around 1926. Thermal noise is a Gaussian process, which means that the probability distribution is as follows:

$$p(b) = \frac{1}{\sigma\sqrt{2\pi}} \exp\left[\frac{(b-\mu)^2}{2\sigma^2}\right], \quad [5.15]$$

where μ is the mean value (zero, if only the noise contribution is considered) and σ is the RMS amplitude, or the standard deviation of the noise.

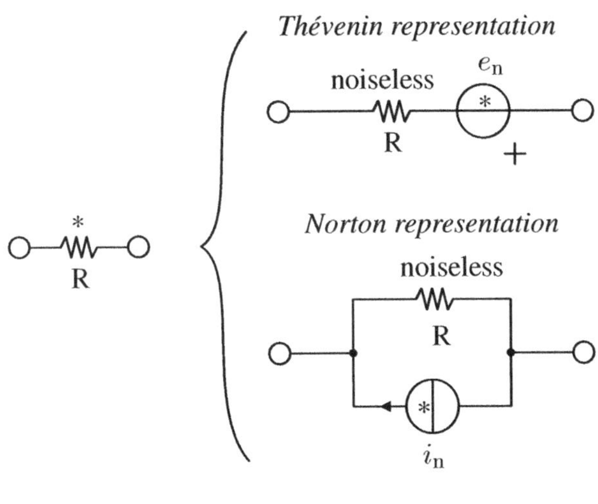

Figure 5.3. *Thévenin and Norton equivalent noise models for a real resistance. Asterisks indicate explicitly which devices are generating noise*

In a resistor R, given the linear relation between voltage and current (the well-known Ohm's law), the noise can be expressed in a perfectly equivalent way as the RMS value of the current or the voltage. For that reason, Figure 5.3 shows the two noise equivalent circuits (Thévenin and Norton) by representing the noise contributions as external sources. Given a frequency range caught between f_1 and $f_2 > f_1$, the square of the RMS value of the two generators can be calculated as follows:

$$\overline{e_{\mathrm{n}}^2} = \int_{f_1}^{f_2} 4k_{\mathrm{B}}TR\mathrm{d}f = 4k_{\mathrm{B}}TR(f_2 - f_1) \qquad [5.16]$$

$$\overline{i_{\mathrm{n}}^2} = \int_{f_1}^{f_2} \frac{4k_{\mathrm{B}}T}{R}\mathrm{d}f = \frac{4k_{\mathrm{B}}T}{R}(f_2 - f_1), \qquad [5.17]$$

where $k_{\mathrm{B}} = 1.38 \times 10^{-23}\mathrm{J/K}$ is the Boltzmann constant, T is the temperature in kelvin and R is the resistance. The spectral densities being constant, the thermal noise is a white noise. To have an order of magnitude in mind, the voltage noise contribution of a $1\,\mathrm{k\Omega}$ resistance at $300\,\mathrm{K}$ is approximatively $4\,\mathrm{nV}/\sqrt{\mathrm{Hz}}$ and the corresponding current noise is $4\,\mathrm{pA}/\sqrt{\mathrm{Hz}}$.

As in equations [5.16] and [5.17], a high value of resistance generates a relatively high voltage noise and a low current noise for the same temperature. Conversely, a low value of resistance gives a low voltage noise and high current noise. The value of the resistance can hence be designed to minimize the thermal noise contribution, if the most detrimental contribution in the circuit is recognized.

Another possibility to reduce the thermal noise is to decrease the temperature T, but this is the last resort after a skillful and comprehensive analysis has been done.

5.3.2. *Flicker or* $1/f$ *noise*

This noise contribution is characterized by a power spectral density that has a $1/f$ dependence. It is therefore a "pink" noise that can be quite relevant for low-frequency and very low frequency circuits. Depending on the cases, its RMS value can be expressed as a voltage or a current contribution:

$$\overline{e_{\mathrm{n}}^2} = \int_{f_1}^{f_2} \frac{K_{\mathrm{e}}^2}{f} \mathrm{d}f \qquad [5.18]$$

$$\overline{i_{\mathrm{n}}^2} = \int_{f_1}^{f_2} \frac{K_{\mathrm{i}}^2}{f} \mathrm{d}f, \qquad [5.19]$$

where K_{e} and K_{i} are constant. Flicker noise tends to sneak into a surprisingly vast amount of disciplines [PRE 78].

This noise contribution is usually found in active devices. In some cases, especially with very old carbon composition types, even resistors tend to exhibit a certain degree of flicker noise. In this context, this noise contribution is often referred to as “excess noise”, as something that adds up to the intrinsic thermal noise discussed in section 5.3.1. Modern good quality metal-film resistors tend to exhibit very low excess noise.

5.3.3. *Avalanche or breakdown noise*

When a reversely biased PN junction in a diode or in a transistor is put in avalanche, there is a noise that is originated by the random collision of carriers during the avalanche process. It is a noise term that needs to be taken into account in some cases with Zener diodes, but also with avalanche photodiodes (where the avalanche phenomenon is exploited to amplify carriers), as well as photomultipliers.

5.3.4. *Burst or “popcorn” or random telegraph signal noise*

At the output of some active devices (such as an operational amplifier), some very fast shifts of the average level might appear randomly, producing a sound of small explosions when amplified and heard through a loudspeaker. Also called “popcorn noise,” or “random telegraph signal noise”, for this reason, it is related to fabrication defects, especially during clean room processes. Observed with the oscilloscope, the popcorn noise shows a distinct signature. Modern discrete operational amplifiers sold nowadays are practically exempt from this noise contribution, but modern transistors built with advanced submicron processed are still affected by it, as shown in Figure 5.4.

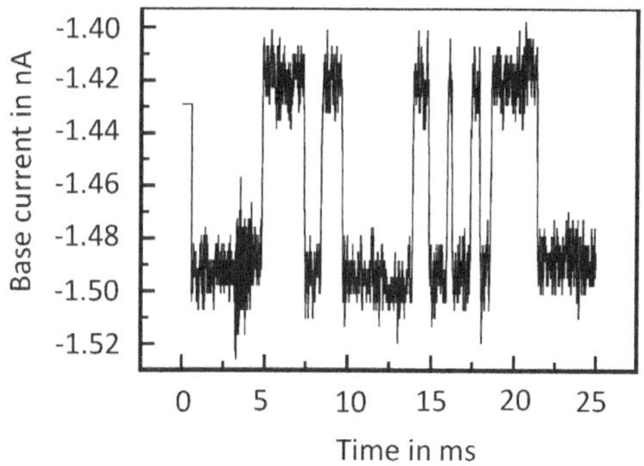

Figure 5.4. *Artifacts of the random telegraph noise on the base current of a submicron heterojunction SiGe/Si bipolar transistor. Courtesy of Mireille Mouis*

5.3.5. *Shot noise or Poisson noise*

A Poisson statistical process corresponds to the following conditions:

– we are counting discrete events occurring in a certain amount of time;

– the events are statistically independent.

Noise with those properties is also called *shot noise* and its probability law is, unsurprinsingly, a Poisson distribution:

$$\mathbb{P}(N_t = k) = \exp(-\lambda t)\frac{(\lambda t)^k}{k!}, \qquad [5.20]$$

where $\mathbb{P}(N_t = k)$ is the probability that we obtain k events in our count. A current is a flow of carriers. In an ordinary conductor, they are electrons bearing a negative charge. In a semiconductor, they might be electrons or holes (positive charges), depending on the doping. Therefore, each time we measure a current we are in practice counting the number of carriers transiting in the unit of time: each pass is a discrete event, hence the presence of noise with a Poisson distribution. One of the most classical examples of Poisson noise appears to be noise in photodiodes. In fact, when carriers have to pass through a potential barrier (such a PN junction), and we may just be in the case corresponding to the previous definition. Its power spectral density is approximatively constant and depends from the average current i_D flowing in the barrier. Therefore, considering a band caught between frequencies f_1 and $f_2 > f_1$, the square of the RMS value of the noise contribution can be calculated as follows:

$$\overline{i_{\text{n}}^2(t)} = \overline{[i(t) - i_D]^2} = \int_{f_1}^{f_2} 2qi_D \text{d}f = 2qi_D(f_2 - f_1) \qquad [5.21]$$

where $i(t)$ is the instantaneous value of the current, q is the charge of the carriers and the notation $\overline{i_{\text{n}}^2(t)}$ indicates the time average of $i_{\text{n}}^2(t)$:

$$\overline{i_{\text{n}}^2(t)} = \lim_{T \to +\infty} \frac{1}{T} \int_{-T/2}^{T/2} i_{\text{n}}^2(t) \text{d}t \qquad [5.22]$$

In practice, in equation [5.21] we are integrating a constant power spectral density (applying the trick seen in section 5.2.3). The shot noise is therefore a *white noise*.

5.4. Limits of modeling

In the previous section, we have adopted some simple models for noise modeling. These present some important convergence problems and, albeit useful, must be handled with care. For example, calculating the power of white noise implies taking into account all the frequency axis, yielding to an integral that does not converge. The white noise model is, therefore, simplified and valid only for frequencies below than a certain limit (around 10^{12} Hz for most conductors).

Regarding pink noise, it is obvious that expressions [5.18] and [5.19] do not converge if $f_1 \to 0$. In other words, the $1/f$ model does not work down to DC. For most situations, working around this problem involves separately specifying in the data sheet the noise of a device as a peak-to-peak value for frequencies up to $10\,\mathrm{Hz}$, and this is done very often for operational amplifiers.

For more demanding applications, such as the long-term determination of the accuracy of reference clocks in metrology applications, more advanced mathematical tools have been developed, such as the Allan variance [ALL 65], but an accurate description is well beyond the scope of this document.

5.5. Contributions from stochastically independent noise sources

Combining contributions of different voltage and current sources is an easy task when dealing with a linear circuit. In fact, the linearity ensures that the superposition term can be applied. When we have different noise sources in a circuit, we are most of the time interested in calculating the statistical properties (such as RMS amplitude) of noise in different nodes of the circuit. If the circuit is linear, the superposition theorem can be applied with all the noise sources. However, certain statistical properties of voltages and currents (and in particular their RMS value) can be easily deduced if the different sources are statistically independent.

Let us consider the circuit shown in Figure 5.5, composed of two resistances in series, at the same absolute temperature T. The instantaneous total noise voltage that can be measured at the output of the circuit is given by the sum of the two instantaneous noise contributions (superposition theorem):

$$e_{\mathrm{t}}(t) = e_{\mathrm{n}1}(t) + e_{\mathrm{n}2}(t). \qquad [5.23]$$

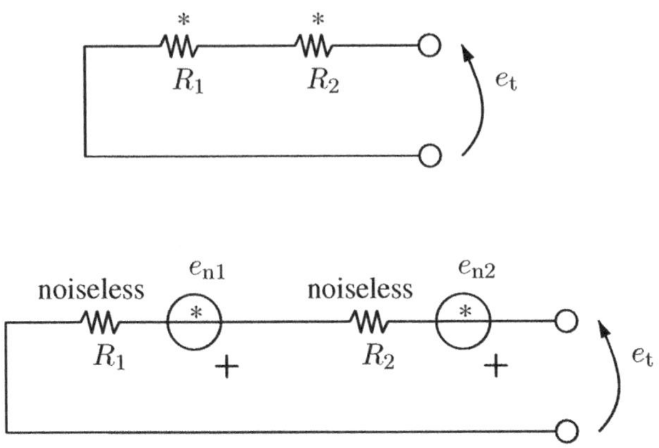

Figure 5.5. *Summing up statistically independent noise contributions: an example with two resistances at the same temperature T*

However, often we can not predict the instantaneous value of the contributions, but only the square of their RMS amplitude as seen in section 5.3.1:

$$e^2_{1,\mathrm{RMS}} = \overline{e^2_1(t)} = 4k_\mathrm{B}TR_1B \qquad [5.24]$$

$$e^2_{2,\mathrm{RMS}} = \overline{e^2_2(t)} = 4k_\mathrm{B}TR_2B \qquad [5.25]$$

where B is the bandwidth of the circuit used for measuring the voltage noise given by the two resistances. It makes sense to see whether we can calculate the RMS value of $e_\mathrm{t}(t)$, and indeed this is possible:

$$\begin{aligned}\overline{e_\mathrm{t}(t)^2} &= \overline{[e_\mathrm{n1}(t)+e_\mathrm{n2}(t)]^2} = \overline{e_\mathrm{n1}(t)^2} + \overline{e_\mathrm{n2}(t)^2} + \overline{2e_\mathrm{n1}(t)e_\mathrm{n2}(t)} \\ &= \overline{e_\mathrm{n1}(t)^2} + \overline{e_\mathrm{n2}(t)^2}, \end{aligned} \qquad [5.26]$$

where the last passage is made possible by the fact that the two noise sources are statistically independent and have a zero average. This equation is often written with RMS amplitudes:

$$e_{\mathrm{t,RMS}} = \sqrt{e^2_{1,\mathrm{RMS}} + e^2_{2,\mathrm{RMS}}}. \qquad [5.27]$$

So the RMS amplitude of the resulting noise contribution is equal to square root of the sum of the square of RMS amplitudes of the two contributions taken separately. It is worth stressing once again that this results is a direct consequence of the linearity and the superposition theorem, and is easily generalizable to any number of statistically independent sources. It *should not* be used when the noise sources have a certain degree of correlation or when the nonlinearity of the circuit makes sort that the superposition theorem cannot be applied.

5.6. Noise equivalent bandwidth and noise factor

The expressions that we have adopted to calculate RMS amplitudes in the previous sections are quite often related to the integration of the power spectral density (or a quantity proportional to it) over a certain frequency range. Let us consider the situation shown in Figure 5.6, where a white noise is filtered by an ideal low-pass filter. In this case, it is particularly easy to calculate the overall power, which is carried by the signal after the filtering process:

$$P_{\mathrm{ideal,B}} = \int_0^B H_0^2 A \mathrm{d}f = H_0^2 AB. \qquad [5.28]$$

In practice, as we have seen in Chapter 3, an ideal filter cannot be built. Therefore, the filtering process requires the calculation of an indefinite integral, as shown graphically in Figure 5.6:

$$P_{\mathrm{real}} = \int_0^{+\infty} |H(f)|^2 A \mathrm{d}f. \qquad [5.29]$$

A useful concept is the equivalent noise band, which is the *band* B_{eq} *of an ideal filter, whose output power is exactly the same as the output power of the real filter when the same white noise is fed at the input of the two filters*. Combining equations [5.28] and [5.29], we obtain:

$$B_{\mathrm{eq}} = \int_0^{+\infty} \frac{|H(f)|^2}{H_0^2} \mathrm{d}f. \qquad [5.30]$$

The input signal being a white noise, this is a way to characterize the filter and simplify calculations.

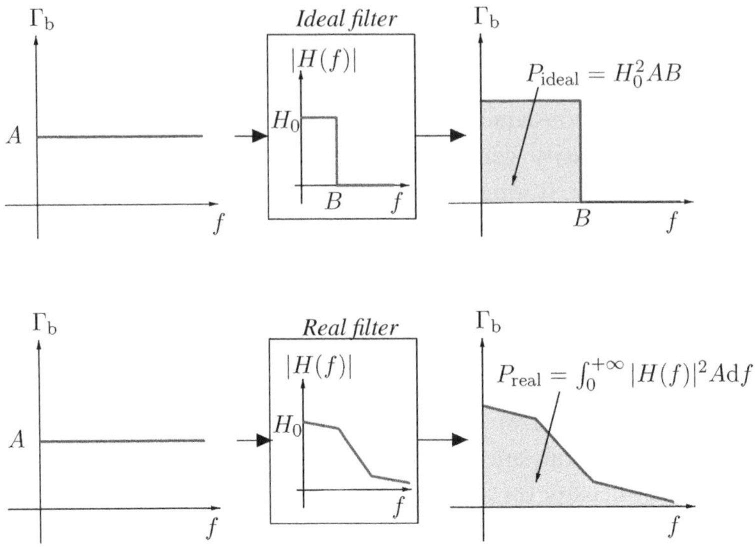

Figure 5.6. *Graphical representation of the total average power at the output of an ideal and a real filter, fed by the same white noise*

As an example, let us take a transfer function of a first-order low pass filter and calculate its equivalent noise bandwidth:

$$H(f) = \frac{1}{1 + jf/f_0}. \tag{5.31}$$

Recalling the definition given above:

$$\begin{aligned} B_{eq} &= \int_0^{+\infty} \left| \frac{1}{1+jf/f_0} \right|^2 df \\ &= \int_0^{+\infty} \frac{1}{1+(f/f_0)^2} df = f_0 \frac{\pi}{2} \approx 1.57 f_0. \end{aligned} \tag{5.32}$$

Another concept that is often employed for a two-port network (for example an amplifier) is the *noise figure*. It quantifies the degradation of the quality of the signal by accounting for the loss in the SNR implied in the process.

$$F = \left. \frac{S_{\mathrm{NR,input}}}{S_{\mathrm{NR,output}}} \right|_{\mathrm{lin}} = \frac{S_{\mathrm{I}}/N_{\mathrm{I}}}{GS_{\mathrm{I}}/[G(N_{\mathrm{I}}+N_{\mathrm{A}})]} = 1 + \frac{N_{\mathrm{A}}}{N_{\mathrm{I}}} \qquad [5.33]$$

where S_{I} is the power of the input signal, N_{i} is the power of the input noise andN_{A} is the additional noise introduced by the network as well as G, the power gain (which can be less than 1 if the network attenuates the signal). We are working here with the powers of signal and noise, which are summed together. Thus, we are supposing that they are statistically independent, as seen in section 5.5. Very often, the noise figure is expressed in dB:

$$N_{\mathrm{F}} = 10\log_{10}(F) = 10\log_{10}\left(1 + \frac{N_{\mathrm{A}}}{N_{\mathrm{I}}}\right) \qquad [5.34]$$

Of course, dealing with powers, G in this context is the power gain that is not to be confused with the voltage gains we used so often in Chapter 2. When the impedance at the input and output of the amplifiers is the same, the power gain is just the square of the voltage gain, so the conversion is not really a big deal.

Equation [5.33] tells that the noise figure of a two-port network depends both on the noise introduced by the amplifier itself, and on the noise normally present on the input signal in the best possible conditions.

5.7. Amplifiers and noise

5.7.1. *Noise models of operational amplifiers*

A problem that arises frequently in low noise circuit design is to model the noise contribution of all circuit elements, so as to evaluate the overall noise level at the output of the circuit. The insight gained by this operation allows us to understand what might be changed or optimized in the circuit so that the performances can be increased. We discussed about the thermal noise issued by resistors in section 5.3.1. Ideal inductors and capacitors do not generate noise (but the parasitic resistance of real ones might give its contribution). We see now which models can be adopted for operational amplifiers.

In fact, operational amplifiers are quite complex devices and the detailed description of the noise origin inside them is outside the scope of this introductory document. However, when an operational amplifier is used in a discrete circuit, the designer does not need to know all the intricate internal working details. In the same spirit, a simplified noise model for operational amplifiers is very often specified in the data sheets, as shown in Figure 5.7.

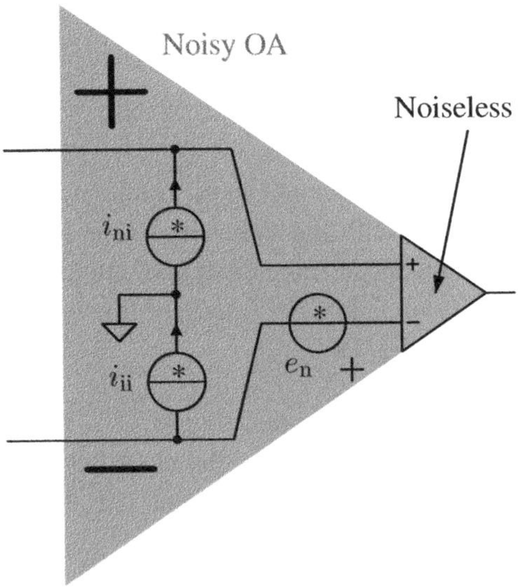

Figure 5.7. *Noise model of an operational amplifier. Three statistically independent noise generators represent the input-referred noise contributions generated by the whole amplifier*

Three statistically independent noise contributions are usually considered:

– e_n is the input voltage noise;

– i_{ni} the input current noise, for the non-inverting input;

– i_{ii} the input current noise, for the inverting input.

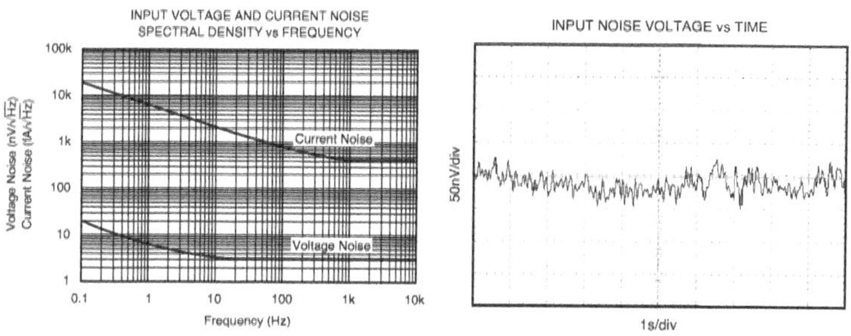

Figure 5.8. *Noise "voltage" and "current" power spectral densities (in the sense of what seen in section 5.2.2), and a recording of several seconds of the output noise voltage. From the data sheet of the OPA227 operational amplifier. Courtesy of Texas Instruments*

The e_n noise generator can be alternatively placed on the inverting input without changing anything on the results (its phase does not matter for any RMS amplitude). Very often, characteristics of i_{ni} and i_{ii} are extremely similar, so that most data sheets only refer to a generic "input-referred current noise" to indicate both of them. Typically, a graph such as the one shown in Figure 5.8 is given, where the spectral densities of e_n and i_{ni}, i_{ii} are shown at a certain frequency range. As it usually happens, the behavior of the plot shows a flicker noise $1/f$ contribution until a certain corner frequency, where flicker noise becomes negligible against thermal noise. It is interesting to remark that the frequency at which this happens is not the same for the voltage noise (around 10 Hz) and the current noise (around 1 kHz). It is also worth noticing that chopper-stabilized amplifiers (see section 2.3.2) are virtually immune from this flicker noise contribution. Note the measurement units employed, which follow the conventions seen in section 5.2.2. A recording of several seconds of the output voltage is often shown in the data sheet: this is usually done to show that the popcorn noise does not appear (compare with Figure 5.4). It is also useful to give an idea on the noise amplitude for very low frequencies, where reasoning on the power spectral density would be meaningless because of the limitations of the models discussed in section 5.4. In fact, the data sheet of the OPA227 also gives an indication of $90\,\text{pV}$ for the typical peak-to-peak input voltage noise in the band $0.1 - 10\,\text{Hz}$.

5.7.2. *Example: noise factor of a non-inverting amplifier*

Let us consider the situation shown in Figure 5.9 on the left, where a non-inverting amplifier is employed to amplify signals coming from a dynamic microphone. We will employ the OPA227 operational amplifier, whose characteristics are shown in Figure 5.8 (a similar analysis is done in [KAR 03] with different devices). The voltage gain of the amplifier is given by:

$$H_0 = -\frac{R_F}{R_G} \qquad [5.35]$$

which corresponds to a power gain:

$$G = \frac{R_F^2}{R_G^2} \qquad [5.36]$$

A very important question to treat, especially if the microphone is expected to sense very delicate sounds, concerns the noise performances of such a system. We show here how a parameter such as the noise factor can be calculated in such a situation. Figure 5.9 on the right shows the noise models of the components employed here: the operational amplifier with its current and voltage noise sources as well as the resistances with their thermal noise. Resistances R_G and R_F determine the input resistance as well as the voltage gain. Resistance R_T is there to follow the old rule of thumb that "an operational amplifier should see the more or less the same resistance from each one of its inputs" to minimize the offsets. We will see if that is justified here.

The determination of the noise factor in such a situation starts with an evaluation of the noise associated with the source, in our case the microphone. Even in a perfectly silent environment, the microphone delivers a noise due at least to the ohmic resistance of its coil, which is represented by means of the e_s voltage source. The square of the RMS voltage of the noise generated by the microphone is thus:

$$\overline{e_s^2} = 4k_B T R_s B \qquad [5.37]$$

where $T = 300\,\text{K}$ is the temperature of the microphone, $R_\text{s} = 250\,\Omega$ is the coil resistance and B is the equivalent bandwidth to consider, from 20 Hz to 20 kHz in our case. However, the total noise voltage present at the input of the amplifier will be lower than e_S, since we have to consider the loading effect of the input impedance of the amplifier (which is equal to R_G). Therefore, at the input of the amplifier, the noise contribution becomes:

$$\overline{e_\text{I}^2} = \left(\frac{R_\text{G}}{R_\text{G} + R_\text{s}}\right)^2 \overline{e_\text{S}^2} \qquad [5.38]$$

To calculate the noise figure of the amplifier, we have to sum up all the contributions of the noise sources shown in Figure 5.9 to calculate a single term of equivalent input noise. This means the total amount of noise injected at the input of a noiseless amplifier, which would produce exactly the same output noise power obtained in our amplifier.

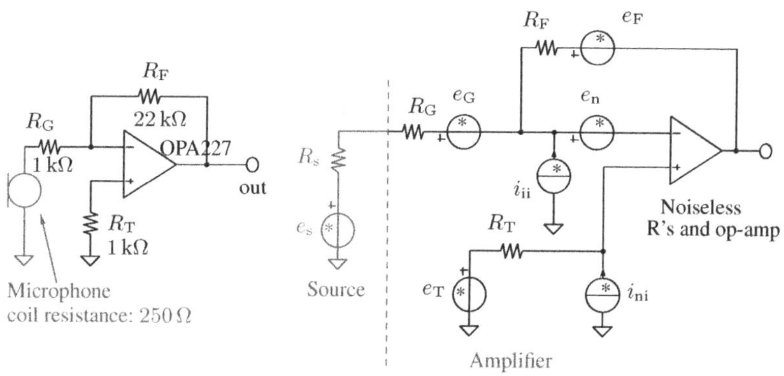

Figure 5.9. *On the left, a non-inverting amplifier amplifying the signal produced by a dynamic microphone. On the right, the same circuit where all the noise sources have been explicitly indicated (therefore, all the components are noiseless)*

Since all noise sources shown in Figure 5.9 are uncorrelated, applying what was described in section 5.5, we expect that the equivalent input noise voltage would be given by a sum of the square of the RMS amplitude of all noise sources with appropriate coefficients, therefore something like:

$$\overline{e_{\mathrm{A}}^{2}} = c_1\overline{e_{\mathrm{n}}^{2}} + c_2\overline{i_{\mathrm{ni}}^{2}} + c_3\overline{i_{\mathrm{ii}}^{2}} + c_4\overline{e_{\mathrm{T}}^{2}} + c_5\overline{e_{\mathrm{G}}^{2}} + c_6\overline{e_{\mathrm{F}}^{2}} \qquad [5.39]$$

where c_2 and c_3 have the dimension of a resistance squared and where other constants are dimensionless. The circuit being linear, each source can be considered separately.

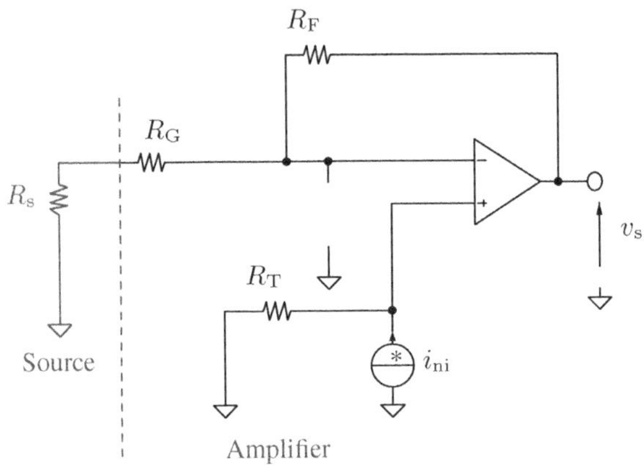

Figure 5.10. *Calculating the c_2 coefficient by switching off all the noise sources except i_{ni}*

As an example, let us calculate c_2, which represents the influence of the current source associated with the non-inverting input. Probably, the most straightforward way to proceed is to calculate the output noise voltage and then divide it by the voltage gain of the circuit to calculate the equivalent input noise contribution. Therefore, we consider the circuit shown in Figure 5.10 where only the noise source i_{ni} is kept active. Note that we keep in place the resistance R_{s}, which represents the microphone, even if the associated noise source is switched off. With a little effort of circuit analysis, it is not difficult to calculate the instantaneous voltage v_{s} *at the output* of the circuit by supposing to know the value of the current i_{ni}:

$$v_{\mathrm{s}} = i_{\mathrm{ni}} R_{\mathrm{T}} \left(1 + \frac{R_{\mathrm{F}}}{R_{\mathrm{s}} + R_{\mathrm{g}}}\right) \qquad [5.40]$$

and finally divide that for the voltage gain calculated in equation [5.35] to obtain the equivalent input noise contribution. Since in equation [5.39] what matters is the square of the RMS voltage, we obtain the equivalent noise contribution associated with the current noise of the non-inverting input of the op-amp:

$$c_2\overline{i_{\rm ni}^2} = \left(\frac{R_{\rm G}}{R_{\rm F}}\right)^2 R_{\rm T}^2 \left(1 + \frac{R_{\rm F}}{R_{\rm s}+R_{\rm g}}\right)^2 \overline{i_{\rm ni}^2} \qquad [5.41]$$

Note how the choice of the sign of the source $i_{\rm ni}$ done in the model shown in Figure 5.9 does not affect the final result and it is therefore purely conventional. This is true for all noise sources.

Term	Coefficient	Source	Sp. density	Contribution
c_1	$\left(\frac{R_{\rm G}}{R_{\rm F}} + \frac{R_{\rm G}}{R_{\rm s}+R_{\rm G}}\right)^2$	$e_{\rm ni}$	$3\,{\rm nV}/\sqrt{\rm Hz}$	$(0.36\,\mu{\rm V})^2$
c_2	$\left(\frac{R_{\rm G}}{R_{\rm F}}\right)^2 R_{\rm T}^2 \left(1+\frac{R_{\rm F}}{R_{\rm s}+R_{\rm g}}\right)^2$	$i_{\rm ni}$	$< 1.5\,{\rm pA}/\sqrt{\rm Hz}$	$(0.18\,\mu{\rm V})^2$
c_3	$R_{\rm G}^2$	$i_{\rm ii}$	$< 1.5\,{\rm pA}/\sqrt{\rm Hz}$	$(0.21\,\mu{\rm V})^2$
c_4	$\left(\frac{R_{\rm G}}{F_{\rm F}} + \frac{R_{\rm G}}{R_{\rm G}+R_{\rm s}}\right)^2$	$e_{\rm T}$	$4\,{\rm nV}/\sqrt{\rm Hz}$	$(0.49\,\mu{\rm V})$
c_5	$\left(\frac{R_{\rm G}}{R_{\rm G}+R_{\rm s}}\right)^2$	$e_{\rm G}$	$4\,{\rm nV}/\sqrt{\rm Hz}$	$(0.46\,\mu{\rm V})^2$
c_6	$\left(\frac{R_{\rm G}}{R_{\rm F}}\right)^2$	$e_{\rm F}$	$19\,{\rm nV}/\sqrt{\rm Hz}$	$(0.12\,\mu{\rm V})^2$
Grand total		$e_{\rm A}$		$(0.82\,\mu{\rm V})^2$
Input noise	$\left(\frac{R_{\rm G}}{R_{\rm G}+R_{\rm s}}\right)^2$	$e_{\rm s}$	$2\,{\rm nV}/\sqrt{\rm Hz}$	$(0.23\,\mu{\rm V})^2$
Noise figure				$11.3\,{\rm dB}$

Table 5.1. *Terms in equation [5.39] for the various noise sources, integrated in the band B between $20\,{\rm Hz}$ and $20\,{\rm kHz}$ with an ambient temperature $T = 300\,{\rm K}$*

Similar calculations can be repeated for all coefficients and the results are summarized in Table 5.1. We calculated there the noise spectral density for each source and, in the last column the total contribution to the terms of equation [5.39], integrated in the audio band. The noise figure is then calculated by means of equation [5.34].

Such a breakdown analysis is interesting since it tells us the following things:

– the most relevant term in equation [5.39] comes from resistance R_T, which creates a lot of noise. Given the low values of R_G and R_F, the tiny offset generated by the bias currents of the operational amplifier will be probably a negligible nuisance, if compared with the benefits of removing that source of noise. Putting $R_\mathrm{T} = 0$ also eliminates the c_2 coefficient, therefore further reducing the noise produced by the amplifier, yielding a noise figure of $9.3\,\mathrm{dB}$;

– the component that taken alone produces the highest spectral density of noise is by far resistance R_F. However, its contribution to the noise produced by the amplifier is the smallest, due to the place which occupies in the circuit;

– the current noise is estimated in a very pessimistic way since $1.5\,\mathrm{pA}/\sqrt{\mathrm{Hz}}$ is the spectral density at a frequency of $20\,\mathrm{Hz}$, where the $1/f$ noise greatly affects the performances. However, the terms associated with current noise are already relatively small in the table so that they do not deserve a more detailed analysis at least at this step. When needed, one may combine a model such as the integrand of equation [5.18] with a white noise model to represent the graph in Figure 5.8 with the appropriate coefficients and then integrate the resulting expression.

Overall, a noise figure of $11.3\,\mathrm{dB}$, which can be reduced to $9.3\,\mathrm{dB}$ by shorting R_T, is a pretty decent result in this situation, since a real microphone will probably deliver a noise voltage larger than the thermal contribution of its coil (after all, there is an acoustic noise level which will be recorded). However, we notice that the most relevant contribution will be given by the resistances and not by the operational amplifier. Unfortunately, the data sheet of the OP227 only specifies the typical noise performances and not the maximum ones. Therefore, the noise figure calculated here will be "typical" as well: this automatically rules out the choice of the OP227 in those situations where noise performances must be guaranteed.

5.7.3. *Noise models of instrumentation amplifiers*

As we saw in Chapter 2, different structures of instrumentation amplifiers exist. Therefore, there are slightly different noise models that can be applied for them. For the most classical structure of instrumentation amplifier discussed in section 2.4.4, the usual noise model is shown in Figure 5.11. Similarly to the operational amplifier, we have three noise sources e_{ni}, i_{ii} and i_{ni} representing the noise contributions referred to the input, exactly as we saw for operational amplifiers in section 5.7.1. On the one hand, there is also (usually) the contribution of a fourth voltage noise source e_{no} referred to the output. When the gain is relatively low, the output noise of the device tends to dominate. On the other hand, when the gain is high, it is the input-referred noise that plays an important role. The origin of the input and output noise terms are easy to understand by studying the structure of the instrumentation amplifier shown in Figure 2.7. In fact, the instrumentation amplifier is divided into a differential input stage, which provides the buffering and most part of the differential gain, and an output stage, which is a differential amplifier with a fixed gain. This is also reflected by the noise contributions. Resistance R_{g} also introduces an additional noise term (not shown in the circuit), whose influence depends on the exact internal structure of the amplifier.

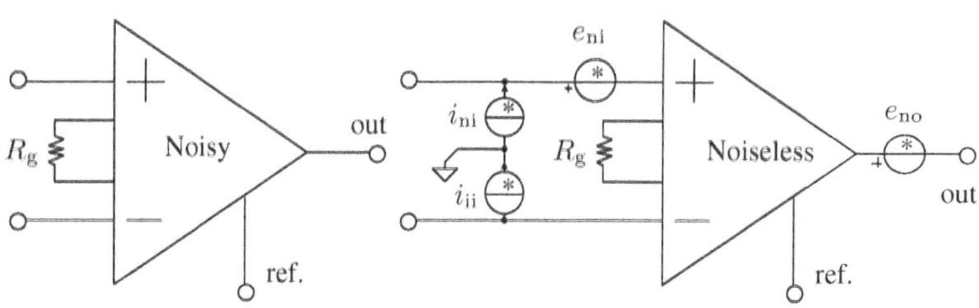

Figure 5.11. *Typical noise model of a 3 op-amps instrumentation amplifier*

5.8. Noise from "outer space": electromagnetic compatibility

When deploying an electronic circuit in a real-world situation, we sometimes must face quite bad surprises. It often happens that two circuits separately working perfectly on the bench of a laboratory cease to work when put in close proximity. Often, a real-life environment is quite noisy from the electromagnetic point of view, from perturbations coming from mains frequency, Radio Frequency (RF) devices and so on [CHA 06]. Problems like those are related to the interaction of a circuit with its "external" environment. In the case of a sensitive measuring device, great care is to be devoted to avoid that external noise sources perturb its most delicate sections. Potentially noisy appliances should be screened in an efficient way to avoid perturbing devices put in their proximity.

All this and much more constitute a realm called Electromagnetic Compatability (EMC). Suspicions of black magic are sometimes raised against experts of this domain, probably because EMC contains elements of low-frequency electronics, but also wave propagation and RF design. Many issues arise from coupling between separate circuits in which unwanted or uncontrolled emission perturbs in some extent a victim circuit. A distinction can be done between short-range low-frequency coupling, where it is possible to identify and separate contributions from electric and magnetic fields, or high-frequency coupling, where the two are indissociable. The distinction in this context is related to the size of the circuit (represented by a characteristic length l) and the wavelength λ of the perturbing signal. If λ is smaller or comparable to l, then we are in a high-frequency framework. If λ is much greater than l, we are in the low-frequency case.

Table 5.2 shows some of the most frequent causes of problems and suggests how to possibly fix them. If the causes are clear in the schematics in the real-world situations where they can sneak in some quite subtle ways. Here, a small list is as follows:

– (A) *Common impedance coupling*: It is a cause of conductive coupling due to the fact that there is no such thing as a zero impedance. So if the source and the victim circuits share a common conductor, the voltage drop on its impedance Z due to the current i_s is different from zero and adds up to the sensor voltage. This arises very often with reference or ground connections, hence the classical suggestion of keeping separate references of different circuits, putting them together in a single point (star connection).

– (B) *Inductive coupling*: At low frequency, a variable magnetic field due to currents loops in the source circuit induces a voltage in the victim source. The field and the voltage being proportional to the total area of the loops, a first solution is to try to reduce them as much as possible. Twisted cables are also employed so that magnetic couplings tend to cancel out due to the twisting. This solution is employed for Ethernet and telephone cables. When twisting is not possible, screening can be effective with high-permeability materials (mu-metal).

– (C) *Capacitive coupling*: At low frequency, parasitic capacitive effects due to the electric field between two conductors can perturb a low-level signal. A first solution is to adopt an electrostatic shield or guard. A second possibility is to employ a differential signal where the information is conveyed by the voltage difference between two conductors that are kept as physically close as possible. If the same perturbation is therefore coupled in both of them in the same amount, a differential amplifier will be able to cancel out almost completely its influence, thus retrieving the signal of interest.

– (D) *Radiative coupling*: At high frequency, the source leaks (or emits on purpose) a certain RF power that is absorbed by the victim circuit. Injection of sufficient RF power in an unprotected low-frequency circuit is a guaranteed way to make it behave in a strange way. Everyone has probably heard the funny sounds emitted by an audio power amplifier when a cellular phone rings nearby. The RF waves emitted by the telephone enter the audio circuits by connection cables as antennas. The high-frequency RF signal then drives the input stages of the audio amplifier to saturation and the resulting nonlinearities perform a sort of envelope detection. The result is that an unpleasant audible signal in the audio range is processed by subsequent stages and sent to the loudspeakers. Solutions are not always easy to be identified but tend to reduce the unwanted RF power emitted (norms exist for that), or filter for RF each signal at the input of a potentially sensitive circuit.

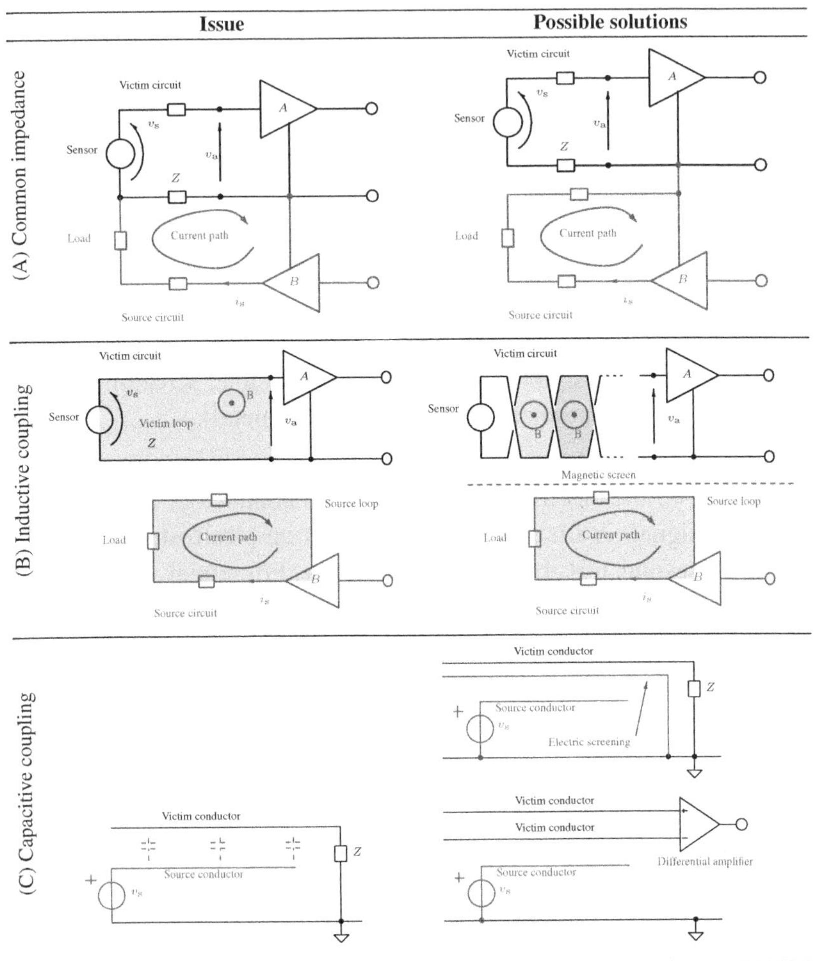

Table 5.2. *Examples of low-frequency coupling between circuits with possible solutions*

Of course, in practice, several EMC issues can arise at the same time and interpreting the physical coupling mechanisms correctly requires a certain degree of experience (here is where all that black magic stuff comes into the picture). The first idea involves trying to make sure that low-frequency and RF emissions are below limits fixed by specific norms. Nowadays, *all* electronic devices put in the market must be thoroughly tested to prove that international norms are respected. The second idea is to make sure that delicate circuits are adequately screened. Since a certain degree of communication with the external world is always necessary, a screening cannot be complete, for example a sheet of metal can have a hole so that a cable can pass. The problem is that a hole of the wrong size in the wrong position might completely remove the effectiveness of an RF screening. From a very practical point of view, here are some suggestions useful when working with low-frequency analog circuits:

– beware of the power supply noise: good operational amplifiers reject quite well a 50 Hz or 60 Hz ripple on their power supply, but not always a 50 kHz noise from a switching power supply, let alone 10 MHz. So, keep analog power supply rails as clean as possible;

– beware of the magnetic coupling at 50 Hz or 60 Hz from mains transformers. Standard ones (with E-I cores) tend to have a certain leak of magnetic flux that can do nasty things. The orientation of the transformer matters and toroidal-core ones tend to be less subject to this problem;

– beware of digital circuits. Fast voltage transitions at the output of logic gates can result in a considerable amount of noise on a wide RF bandwidth. Keep physically separated analog and digital front-ends. In delicate cases, optocouplers, insulation amplifiers and insulated power supply converters can be helpful;

– filter for RF any input of a circuit potentially exposed to RF. Since everyone has a portable phone or a wireless device, this translates to: *always* filter for RF any input of an analog circuit;

– beware of long cables or PCB tracks; we are constantly subjected to magnetic and electric fields at mains frequency. Shield if necessary. Employing differential signals is also an effective option;

– the voltage reference node of our circuit may (or not) be connected to the chassis ground or (if it is metallic) to the box containing the instrument. At its turn, often for safety reasons, the chassis is tied to the Earth connection of power plugs. It is worth remembering that the reference, the chassis and the Earth are three different things. They may or not be at the same potential and if they are tied together there may be some contact resistance to consider. An unwise ground connection is also a classic source of ground loops.

5.9. Conclusion

In this chapter, we briefly discussed the noise sources that can perturb the analog front end of an electronic measurement system. Far from being comprehensive, we have just introduced a brief overview of the principles of stochastic signal modeling and dealt with some types of noise classically found in real-world circuits. We ended with a very concise introduction to the issues of EMC.

Appendix

Legal Notes

IMPORTANT NOTICE

Texas Instruments Incorporated and its subsidiaries (TI) reserve the right to make corrections, enhancements, improvements and other changes to its semiconductor products and services per JESD46, latest issue, and to discontinue any product or service per JESD48, latest issue. Buyers should obtain the latest relevant information before placing orders and should verify that such information is current and complete. All semiconductor products (also referred to herein as "components") are sold subject to TI's terms and conditions of sale supplied at the time of order acknowledgment.

TI warrants performance of its components to the specifications applicable at the time of sale, in accordance with the warranty in TI's terms and conditions of sale of semiconductor products. Testing and other quality control techniques are used to the extent TI deems necessary to support this warranty. Except where mandated by applicable law, testing of all parameters of each component is not necessarily performed.

TI assumes no liability for applications assistance or the design of Buyers' products. Buyers are responsible for their products and applications using TI components. To minimize the risks associated with Buyers' products and applications, Buyers should provide adequate design and operating safeguards.

TI does not warrant or represent that any license, either express or implied, is granted under any patent right, copyright, mask work right, or other intellectual property right relating to any combination, machine, or process in which TI components or services are used. Information published by TI regarding third-party products or services does not constitute a license to use such products or services or a warranty or endorsement thereof. Use of such information may require a license from a third party under the patents or other intellectual property of the third party, or a license from TI under the patents or other intellectual property of TI.

Reproduction of significant portions of TI information in TI data books or data sheets is permissible only if reproduction is without alteration and is accompanied by all associated warranties, conditions, limitations, and notices. TI is not responsible or liable for such altered documentation. Information of third parties may be subject to additional restrictions.

Resale of TI components or services with statements different from or beyond the parameters stated by TI for that component or service voids all express and any implied warranties for the associated TI component or service and is an unfair and deceptive business practice. TI is not responsible or liable for any such statements.

Buyer acknowledges and agrees that it is solely responsible for compliance with all legal, regulatory and safety-related requirements concerning its products, and any use of TI components in its applications, notwithstanding any applications-related information or support that may be provided by TI. Buyer represents and agrees that it has all the necessary expertise to create and implement safeguards which anticipate dangerous consequences of failures, monitor failures and their consequences, lessen the likelihood of failures that might cause harm and take appropriate remedial actions. Buyer will fully indemnify TI and its representatives against any damages arising out of the use of any TI components in safety-critical applications.

In some cases, TI components may be promoted specifically to facilitate safety-related applications. With such components, TI's goal is to help enable customers to design and create their own end-product solutions that meet applicable functional safety standards and requirements. Nonetheless, such components are subject to these terms.

No TI components are authorized for use in FDA Class III (or similar life-critical medical equipment) unless authorized officers of the parties have executed a special agreement specifically governing such use.

Only those TI components which TI has specifically designated as military grade or "enhanced plastic" are designed and intended for use in military/aerospace applications or environments. Buyer acknowledges and agrees that any military or aerospace use of TI components which have ***not*** been so designated is solely at the Buyer's risk, and that Buyer is solely responsible for compliance with all legal and regulatory requirements in connection with such use.

TI has specifically designated certain components as meeting ISO/TS16949 requirements, mainly for automotive use. In any case of use of non-designated products, TI will not be responsible for any failure to meet ISO/TS16949.

Products		**Applications**	
Audio	www.ti.com/audio	Automotive and Transportation	www.ti.com/automotive
Amplifiers	amplifier.ti.com	Communications and Telecom	www.ti.com/communications
Data Converters	dataconverter.ti.com	Computers and Peripherals	www.ti.com/computers
DLP® Products	www.dlp.com	Consumer Electronics	www.ti.com/consumer-apps
DSP	dsp.ti.com	Energy and Lighting	www.ti.com/energy
Clocks and Timers	www.ti.com/clocks	Industrial	www.ti.com/industrial
Interface	interface.ti.com	Medical	www.ti.com/medical
Logic	logic.ti.com	Security	www.ti.com/security
Power Mgmt	power.ti.com	Space, Avionics and Defense	www.ti.com/space-avionics-defense
Microcontrollers	microcontroller.ti.com	Video and Imaging	www.ti.com/video
RFID	www.ti-rfid.com		
OMAP Applications Processors	www.ti.com/omap	**TI E2E Community**	e2e.ti.com
Wireless Connectivity	www.ti.com/wirelessconnectivity		

Mailing Address: Texas Instruments, Post Office Box 655303, Dallas, Texas 75265

Figure A.1. *Notes for Texas Instruments datasheets (pp. 46, 59, 60, 115, 141)*

Bibliography

[ALL 65] ALLAN D.W., "Statistics of atomic frequency standards", *IEEE Proceedings*, vol. 54, no. 2, pp. 221–230, 1965.

[ASC 03] ASCH G. *et al.*, *Acquisition de données, du capteur l'ordinateur*, 2nd edition, Dunod, Paris, 2003.

[ASH 76] ASHCROFT N.W., MERMIN N.D., *Solid State Physics*, Saunders, Philadelphia, 1976.

[BUR 94] BURR-BROWN/TEXAS INSTRUMENTS, Designing photodiode amplifier circuits with OPA128, Application Bulletin, Burr-Brown/Texas Instruments, 1994.

[CAM 05] CAMEZIND H., *Designing Analog Chips,* Virtualbookworm.com Publishing, College Station, 2005.

[CHA 06] CHAROY A., *CEM Parasites et perturbations des électroniques*, vol. 1–4, 2nd edition, Dunod, Paris, 2006.

[EIN 05] EINSTEIN A., "Über einen die erzeugung und verwandlung des lichtes betreffenden heuristischen gesichtspunkt", *Annalen der Physik,* vol. 322, pp. 132–148, 1905.

[FRA 15] FRANCO S., *Design with Operational Amplifiers and Analog Integrated Circuits,* 4th edition, McGraw-Hill, New York, 2015.

[GAR 90] GARDNER W.A., *Introduction to Random Processes with Applications to Signals & Systems,* 2nd edition, McGraw-Hill, New York, 1990.

[HOR 15] HOROWITZ P., HILL W., *The Art of Electronics*, 3rd edition, Cambridge University Press, New York, 2015.

[KAR 00] KARKI J., Signal conditioning piezoelectric sensors, Application Report, Texas Instruments SLOA033A, 2000.

[KAR 03] KARKI J., Calculating noise figure in op amps, *Analog Application Journal Q 4*, Texas Instruments SLTY094, 2003.

[KES 05] KESTER W. (ed.), *The Data Conversion Handbook*, Newnes Elsevier, Burlington, 2005.

[KES 09] KESTER W., ADC Architectures III: Sigma-Delta ADC Basics, Analog Devices MT-022 Tutorial, 2009.

[MAL 15] MALVINO A.P., BATES D., *Electronic Principles*, 8th edition, McGraw-Hill, New York, 2015.

[NIS 90] NIST ITS-90 Thermocouple Database, NIST Standard Reference Database 60, Version 2.0 (Web Version), http://srdata.nist.gov/its90/main/, 1990.

[ORC 66] ORCHARD H.J., "Inductorless filters", *Electronics Letters*, vol. 2, no. 6, pp. 224–225, 1966.

[PRE 78] PRESS W.H., "Flicker noise in astronomy and elsewhere", *Comments Astrophys*, vol. 7, no. 4, pp. 103–119, 1978.

[ROE 75] ROEDEL R., VISWANATHAN C.R., "Reduction of popcorn noise in integrated circuits", *IEEE Transactions on Electron Devices*, vol. 22, no. 10, pp. 962–964, 1975.

[WAN 05] WANG T., ERHMAN B., Compensate transimpedance amplifiers intuitively, Application Report SBOA055A, 1993 (Revised March 2005).

[WIL 85] WILLIAMS J., High speed comparator techniques, Linear Technology Application Note 13, 1985.

[WIL 90] WILLIAMS J., Bridge circuits, marrying gain and balance, Linear Technology, Application Note 43, 1990.

[WIL 11] WILLIAMS J., HOOVER G., Fidelity testing for A->D converters, Linear Technology, Application Note 132, 2011.

[ZVE 67] ZVEREV A.I., *Handbook of Filter Synthesis,* Wiley, New York, 1967.

Index

Printed and bound by CPI Group (UK) Ltd, Croydon, CR0 4YY

15/09/2026

14962343-0001